GIS Implementation
for Water and Wastewater
Treatment Facilities

Prepared by **Implementing Geographic Information Systems**
Task Force of the **Water Environment Federation**

Ronald Miner, *Chair*

Susan Ancel	Greg Gohrband	Manu Patel
Douglas E. Barker	Indranil Goswami	John Przybyla
Charlie Bristol	Thomas Grala	Ceila Rethamel
James J. Bates, III	Greg Hall	Nicole Schmidt
James Cannistra	Richard Hammond	Barbara Schmitz
Lawrence F. Catalano	Marina Havan-Orumieh	Uzair M. Shamsi
Scott Cattran	Chuck Horne	Edward Speer
Prasad V. Chittaluru	Paul Hsuing	Robert Toring
Sean Christian	Ahmed Husain	Thomas M. Walski
John Duckhorn	Paul Koch	Brooks P. Weaver
Marcel P. Dulay	Andy Moore	Todd Wyman
Yazdan Emrani	Sean Myers	
Robert D. Getter	Gary Ostroff	

Under the Direction of the **Municipal Subcommittee** of the **Technical Practice Committee**

2004

Water Environment Federation
601 Wythe Street
Alexandria, VA 22314–1994 USA
www.wef.org

GIS Implementation for Water and Wastewater Treatment Facilities

WEF Manual of Practice No. 26

Water Environment Federation

WEF Press

McGraw-Hill
New York   Chicago   San Francisco   Lisbon   London   Madrid
Mexico City   Milan   New Delhi   San Juan   Seoul   Singapore
Sydney   Toronto

The McGraw-Hill Companies

Cataloging-in-Publication Data is on file with the Library of Congress.

4 5 6 7 8 9 10   IBT/IBT   1 9 8 7 6 5 4 3 2 1

ISBN 0-07-145305-9

*The sponsoring editor for this book was Larry S. Hager and the production supervisor
was Pamela A. Pelton. It was set in Times Roman by WEF Press. The art director for
the cover was Anthony Landi.*

Printed and bound by IBT Global

 This book was printed on recycled, acid-free paper containing
a minimum of 50% recycled, de-inked fiber.

Water Environment Federation

Improving Water Quality for 75 Years

Founded in 1928, the Water Environment Federation (WEF) is a not-for-profit technical and educational organization with members from varied disciplines who work toward the WEF vision of preservation and enhancement of the global water environment. The WEF network includes water quality professionals from 79 Member Associations in over 30 countries.

For information on membership, publications, and conferences, contact

Water Environment Federation
601 Wythe Street
Alexandria, VA 22314-1994 USA
(703) 684-2400
http://www.wef.org

Manuals of Practice
of the Water Environment Federation

The WEF Technical Practice Committee (formerly the Committee on Sewage and Industrial Wastes Practice of the Federation of Sewage and Industrial Wastes Associations) was created by the Federation Board of Control on October 11, 1941. The primary function of the Committee is to originate and produce, through appropriate subcommittees, special publications dealing with technical aspects of the broad interests of the Federation. These publications are intended to provide background information through a review of technical practices and detailed procedures that research and experience have shown to be functional and practical.

**Water Environment Federation Technical Practice Committee
Control Group**

G.T. Daigger, *Chair*

B.G. Jones
M.D. Nelson
A.B. Pincince
T.E. Sadick

Authorized for Publication by the Board of Directors
Water Environment Federation

William Bertera, *Executive Director*

Contents

List of Tables

List of Figures

Preface

This book addresses the growing need for professional guidelines in the development of data, software, and work practices related to spatial information. *Geographic information system* (GIS) is used throughout the text to connote the "system" to which these guidelines of practice apply. Defining the universe of activities that should be considered as part of GIS is an important element of the book. The topics to be addressed in this publication include the following:

- What GIS is now and what it will be 10 years from now,
- How the wastewater/water utility can benefit from GIS,
- Planning for GIS,
- Designing the system,
- Database development,
- Applications development,
- Organizational development and change management,
- Technology and obsolescence management, and
- Related technologies and systems integration.

This manual is primarily intended for managers of water and wastewater municipalities, consultants, city engineers and planners, and industry.

This manual was developed under the direction of Ronald Miner, *Chair*.

Principal authors of the publication are

James Cannistra	(4)
Sean Christian	(1, 2)
Greg Hall	(9)
Richard Hammond	(3)
Marina Havan-Orumieh	(7)
Ahmed Husain	(6)
Barbara Schmitz	(9)
Uzair M. Shamsi	(5)
Robert Toring	(8)

Additional content for Chapter 3 was provided by James J. Bates, III, and John Przybyla. Edward Speer and Chuck Horne were contributing authors to Chapter 5. Desi Alvarez and Gerald E. Greene were contributing authors to Chapter 6.

Authors' and reviewers' efforts were supported by the following organizations:

Advanced Infrastructure Management, Inc., Brea, California
ATS-Chester Engineers, Pittsburgh, Pennsylvania
Black & Veatch, Kansas City, Missouri

Camp Dresser and McKee, Cambridge, Massachusetts; Detroit, Michigan; Bellevue, Washington
CGvL Engineers, Irvine, California
CH2M Hill, Redding, California
City of Baltimore, Baltimore, Maryland
City of Downey, Downey, California
EPCOR Water Services, Edmonton, Alberta
Farnsworth Group, Inc., Denver, Colorado
GBA Master Series, Inc., Lenexa, Kansas
Haestad Methods, Inc., Nanticoke, Pennsylvania
HydroQual, Inc., Mahwah, New Jersey
Louisville Water Company, Louisville, Kentucky
Los Angeles County Districts of Sanitation, Whittier, California
Morgan State University, Baltimore, Maryland
Narragansett Bay Commission, Providence, Rhode Island
Parsons, Austin, Texas
PlanGraphics, Inc., Silver Spring, Maryland
PSB & J, Orlando, Florida
Sanborn Map Company, Colorado Springs, Colorado
Sidhu Associates, Inc., Hunt Valley, Maryland
Utility Automation Integrators, Shorter, Alabama
Westin, Inc., Rancho Cordova, California
Woolpert LLP, Dayton, Ohio
Woodward and Corran, Portland, Maine

GIS Implementation
for Water and Wastewater
Treatment Facilities

Chapter 1
Introduction

WHAT ARE GEOGRAPHIC INFORMATION SYSTEMS?

Simply stated, *geographic information systems* (GIS) are tools for interacting with computerized data by its location on planet earth. In addition, an intelligent map provides a tool for analysis, enhanced decision-making through data visualization, and improved access to information. Although this does not sound too profound, it is the underlying principle for interacting with "spatial data".

Computerized data within GIS has an organizational structure or format known as a database, which is used to describe physical features located on our planet. However, databases are not always bound to physical features. For instance, financial databases are used extensively to store information about business transactions or customer history, which represents "nonspatial" data. In any database, data can be organized in rows and columns, where columns represent unique topics (fields) of information and rows represent individual units (records). An example of this structure is a database that describes trees. In this example, fields would include such things as the trunk diameter, species of tree, or an indication on the health of the tree. Every time a tree is cataloged, a new record is created that describes each tree. In this manner, earthly features such as trees, roads, and lakes can be described and recorded in records and fields of information. Geographic information systems represent features on a map using three graphical elements, points (nodes), lines (arcs), and polygons. In this example, trees would be represented on a GIS map by nodes, roads would be represented by arcs, and lakes would be represented by polygons.

A single database has always been an efficient tool for managing information storage and retrieval. However, database technology's most productive use is in the ability to link several databases together for queries and analyses. Without GIS, databases of information have to be linked to each other through the use of established relationships. Relationships between nonspatial databases are established using identical information within fields of each database. Database management software allows for querying and analyses through several databases by using these established relationships or links. This limits analyses to only those databases that were designed to be integrated. With GIS, interacting with individual databases is made possible through relating these data sets based on their spatial components. In this way, databases that have been assigned to an earthly coordinate system can be related to one another as geographic "layers".

TECHNOLOGICAL HISTORY OF GEOGRAPHIC INFORMATION SYSTEMS

The progression of GIS as a technology follows the proliferation of computers from the 1950s to the present day. Significant milestones that were reached during this time include theories on using computers to correctly represent our planet and mimic our daily needs and routines, the progression of software and hardware development, and the founding of industry support groups.

DATABASE INTEGRATION. Before GIS technology, the success of relating nonspatial computerized databases relied heavily on establishing common fields during their design. An example of this would be to compare a list of customer addresses with customer orders. To relate these databases, they would have to share a common field, such as a field for the customer name. Data compiled here would only be able to be compared with other databases that shared the same common field, typically planned during the design of the databases. Now, say that we want to relate this customer information to another database that includes the location of company distribution centers to optimize deliveries. Because the information regarding the location of distribution centers was not originally designed to relate to the other two databases, it is unlikely that there would be a common field that they share. Moreover, it would be difficult to compare data because both the customer information and distribution centers have no spatial reference to each other. Geographic information systems provide the one common field most data can be referenced by, its location on planet earth. The GIS, therefore, can be used to relate and analyze unrelated information through spatial relationships.

AUTOMATED MAPPING. As a technology, GIS can trace its roots back to the 1950s and 1960s. In 1959, a simple model for applying computer technology to cartography was produced. This system, known as map in–map out, sought to automate mapmaking through the integration of database technology with computer-aided drafting (CAD) systems (Dodge et al., accessed 2003). By doing so, maps could be text annotated automatically through the use of an existing database. This drastically improved a cartographer's ability to recreate standardized maps and led to future GIS applications. Today, GIS technology is an integral part of cartography and mapmaking and remains the single most popular use.

Another milestone was reached in automated mapping (AM) from 1965 to 1967 that led to advances in geocoding technology. *Geocoding* is the process of assigning an earthly coordinate to nonspatial database records using indexing techniques. The most common geocoding method used within GIS is a system-

---

atic comparison of street name, number, and zip code, with an established GIS street layer that contains similar data. In 1967, the Census Bureau advanced geocoding technology as a means for automating the accounting of mail-out and mail-in forms in preparation for the 1970 census (Dodge et al., accessed 2003). In doing so, the Census Bureau developed the technology of indexing data through street addresses using GIS, a technique that is widely used today. Through this effort, the Census Bureau developed the TIGER system, which provides a national standard street index for referencing nonspatial data. Today, TIGER system data are provided free of charge and represent a collection of core GIS database layers (base map) used to build spatial data.

FACILITIES MANAGEMENT. Through the integration of database technology with CAD, a new form of information management was developed. Using the cartographer's new tool of automated text annotation, functionality such as finding database information by interacting with an electronic map was introduced. Users could select a map feature, such as a street or tree, and information from the database about the geographic feature would be displayed on the screen. This computerized application was successful in allowing users quick access to information contained in databases using a familiar interface. This type of application is known as facilities management. Coupled with AM, automated mapping/facilities management (AM/FM) systems are widely used in the water and wastewater industries and provide a means for enhanced information management and decision-making. A common application of AM/FM technology is a computerized maintenance management system (CMMS). The CMMS can communicate information about facilities, construction drawings, scheduled maintenance and condition, and televised footage of pipe inspections through an easy-to-use, map-driven interface. Today, the CMMS is being integrated with accounting and financial planning groups to better manage assets.

SPATIAL ANALYSIS. Working in parallel with AM/FM development, researchers in Canada, Great Britain, and the United States have developed a suite of analytical tools and statistical models that extend GIS capabilities far beyond AM/FM applications. These tools and models coupled with AM/FM technology accurately describe most existing GIS today.

Spatial analysis allows users the ability to perform sophisticated calculations based on topology. An example of this would be to compute population within a service area. Both are databases represented by mapped boundaries that are accurately scaled to an adopted earthly coordinate system. In this example, database records of census tracts containing population statistics can be selected using coordinate geometry. The GIS software determines which census tracts would be contained within a service area boundary and returns a result by selecting the records that meet this criterion. This is an example of a simple database query that was performed spatially. Other more sophisticated uses of spatial analyses include mathematical calculations or statistical modeling that take advantage of the same spatial relationships.

TECHNOLOGICAL REVOLUTIONS. Before the development of the Pentium processor, the availability of GIS software was limited because of the large amount of processing power required when performing sophisticated GIS tasks. Geographic information systems computing was restricted to expensive UNIX workstations, which severely limited the technology's ability to grow, as most users could only afford a few workstations and required special knowledge of UNIX platforms. Well aware of the potential, many GIS software developers produced desktop GIS solutions in anticipation of more powerful personal computers (PCs). After the introduction of Pentium technology, the use of GIS grew exponentially. Because of this, the majority of GIS programs that exist in public and private sectors are approximately at the same level of sophistication, with only a few examples of more mature programs from which to learn.

The second revolution occurred with the commercialization of the Internet. More-productive GIS programs within an organization include applications that are delivered through Internet/Intranet technology. This is true for two reasons: (1) an organization can reach the widest amount of users for the time and money investment and (2) the familiarity of Web-browser technology coupled with well-thought-out designs allow for a user-friendly environment. Geographic information systems software developers have built Web technology that takes advantage of the decentralized nature of the Internet.

Advances in mobile computing are ushering in a third revolution within the GIS industry. Mobile technologies such as handheld PCs are enabling field personnel to access geospatial information quickly and easily. Data collection is also enhanced, with an ability to feed information back to a central repository in real time. Although use of GIS with handheld PCs is somewhat in its infancy, mobile computing has been around for many years. However, with the use of GIS on handheld devices, data collection and remote access to geospatial data will never be the same.

METADATA. *Metadata*, or data about data, communicates information on GIS data sets, which is critical for its proper use. Because GIS is based on a collection of spatially rendered information, metadata becomes the foundation for communicating ideas such as what is the official file to be maintained within the system and how often is the information updated. There are current federal and state standards that govern the format of metadata as well as what information is to be collected and reported. Although these standards are not required by public agencies, they could affect your GIS program if you intend to share information with federal and state geospatial clearinghouses or apply for grants. Metadata are key components in managing a successful enterprise system. That is because there is no other medium that directs users of GIS software to the different data resources available while communicating essential knowledge on the proper use of what is stored. Information such as currentness, accuracy, the name of the individual who creates and maintains the information, file path, and definitions of each field within a database are all communicated through a metadata catalog. In addition, license restrictions and requirements can be communicated, which acts as a form of enforcement. After its deployment, metadata will become a valuable reference for GIS users within the Enterprise system. The catalog also establishes a means for promoting what information is available for use. By understanding what resources are available, GIS will be popular tools of choice for future projects. As use of metadata information proliferates, the knowledge base will also experience rapid growth as GIS users look to acquire additional GIS resources in support of individual projects.

PRESENT DAY GEOGRAPHIC INFORMATION SYSTEMS INDUSTRY

Today's GIS industry is a fast-moving and dynamic enterprise. It would have been hard to imagine all of the new and wonderful developments in GIS just five years ago, let alone ten years ago. Yet, we are just beginning to realize the tremendous potential of this powerful tool.

MAPPING. With the advances we have seen in technology in all areas of science and engineering, as well as in GIS, we have experienced a rise in the use of the term "mapping" for describing tasks. Scientists are mapping the human DNA in the genome project; business professionals are mapping their company's strategic plans; and, closer to home, most people are mapping their financial future. Most of these references use the term mapping as a metaphor to give better depth to the specific action taking place, that is, developing a strategic plan or a personal financial plan.

More often than not, the term mapping is used as a visual picture of an area with certain features and attributes such as a site-grading plan or street-improvement map. In general, maps can be divided into two categories, namely *flat maps* and *relational or spatial maps*. Flat maps are typically CAD drawings that depict a specific scenario such as a site plan or plan and profiles of a wastewater pipeline. These maps are simply graphical renderings and do not possess any relational information. For example, a CAD drawing of a wastewater line may have a notation that shows the line to be 0.3 m (12 in.) in diameter and having a length of 90 m (300 ft) from manhole to manhole. However, information regarding the length and the diameter of the line is not stored in any database and is simply a notation on the drawing. A GIS or spatial map, by contrast, stores the attribute information of the same pipe in a database, which is connected to the drawing thereby giving the drawing its "spatial" element. The main difference between a CAD map and a spatial map is that a CAD map is a rendering of a drawing that is stored exactly "as it appears on the screen". A spatial map, by contrast, does not store the map in a conventional sense but rather gives the user flexibility and options in uses of attribute information stored in the database as a means to generate different maps.

AUTOMATED MAPPING/FACILITIES MANAGEMENT. Automated mapping/facilities management is a collection of software, procedures, and policies to help assist in the management of a geographically distributed facility. Automated mapping/facilities management became popular in the 1980s and early 1990s when user-friendly, custom applications were developed and tailored to the needs of markets such as utilities, transportation, and local governments.

At first, AM/FM systems were developed to record asset locations, followed by basic modeling of real-world networks for analysis and planning plus a database to store attributes about assets. In today's environment, agencies are branching out beyond the inventory and map placement of assets and facilities needed for tracking. They need additional layers of information, such as customers and operational performance of their networks, presented graphically to help in the decision-making process.

When GIS is added to the information technology infrastructure of a utility, it enhances the decision-making process of AM/FM and many other utility functions. Spatial data engines sitting atop a standard relational database management system help manage utility data, much of which has a spatial orientation. The current and future trend in AM/FM/GIS is development across departments and using the Internet as a backbone for a Web-enabled, enterprise-wide system.

COMPUTERIZED MAINTENANCE MANAGEMENT SYSTEMS. Computerized maintenance management systems are used to track and address unscheduled or emergency repairs and preventive maintenance of collection or distribution systems. That is why at the heart of every good CMMS is a great workorder management system. A CMMS is used as a database to store all temporal information regarding ongoing operations and maintenance of infrastructure systems such as wastewater, water, and stormdrain pipelines. It can also be used to track the operation and maintenance of pumping stations, water purification plants, and wastewater treatment plants. A CMMS comes in all shapes and varieties, but generally can be classified into two main categories, namely

(1) A CMMS can serve as a stand-alone, nongraphic database of a system's features, attributes, and workorder history or
(2) A more useful application of CMMS is through seamless integration with GIS whereby data can be viewed, accessed, and analyzed easily and effortlessly combining the powerful database capabilities of a CMMS with the graphic and spatial capabilities of a GIS.

The effective use of CMMS is typically preceded by a data-conversion project involving data collection and population for desired infrastructure system attributes.

Depending on the type of system and vendor, CMMS can also include features such as inventory control, capital improvement planning, and automated scheduling of preventive maintenance.

INTEGRATED INFORMATION MANAGEMENT SYSTEMS. A GIS-centric integrated information management system (IIMS) uses GIS at its core to integrate the functions of various departments as well as interface with different software to produce an effective IIMS. Many IIMS take advantage of an existing record-management system within an organization by integrating access to records through a map-driven application. This extends the functionality of both records management and AM/FM systems. An example of this would be an application that can locate documents within a records-management system by selecting a pipe or manhole within a GIS front-end application.

The Internet, advanced computing technologies, and digital information have altered the information landscape. With improved access to data, information and knowledge are no longer time- and place-dependent and new opportunities are emerging to improve all elements of engineering and infrastructure management. To benefit from these advances, municipalities and public works organizations must (1) seamlessly integrate their own digital information resources with relevant information obtained from external sources and (2) bring digital information to their organization and staff as well as the general public in a way that supports sound decisions and effective action.

An effective IIMS can plan, design, test, and deploy systems and techniques for integrating data, information, and knowledge resources into a comprehensive networked information management system.

There are many different types of IIMS, such as in risk management, healthcare, and security systems. Regardless of the specific application of an IIMS, there are certain characteristics that are common to all IIMS

- A management framework,
- Policy and data standards,
- Spatial and tabular data,
- Custom application for data access and use,
- Proper hardware and software, and
- Knowledgeable and trained people.

ENTERPRISE GEOGRAPHIC INFORMATION SYSTEMS. How many people in the public sector have encountered a situation whereby different departments within an organization seem to have different goals and objectives, that at first glance seem almost at odds with each other? This scenario becomes even more confusing when different departments have used different software and consultants to achieve their individual GIS needs and objectives. How many people have wished that all departments within an organization could reach across departmental lines and boundaries and work together in implementing a clear and concise set of goals and objectives that address the primary needs of an organization? Furthermore, wouldn't it be wonderful if this could be achieved by implementing a uniform and seamless GIS? The first question, in essence, defines the problem faced by many public agencies today and the second question defines the primary reason for developing an enterprise-wide GIS.

Increasingly, public agencies are looking at streamlining their individualized GIS operations and moving towards an enterprise-wide GIS. In today's world of shrinking budgets and tight job markets, it is no longer feasible to have separate entities within an organization having separate and different GIS operations. It also makes good sense to achieve efficiency and maximize the dollars invested through achieving the main objective for the whole organization and across many departmental and group lines.

Typically, an enterprise-wide system looks at the needs, objectives, and budget of all departments and units and how individual needs and objectives tie into the bigger picture for the whole organization. This is primarily achieved through a series of meetings, interviews, and discussions with the main stakeholders from different entities. The enterprisewide GIS also looks at background data, studies, master plans, etc., for different divisions. Next, a synergistic enterprisewide plan is developed to achieve economies of scale for common needs such as hardware; software; data-conversion services; GIS applications; metadata standards; and, if needed, consulting services.

PROJECT VISUALIZATION. The nature of any GIS is to model the world around us and describe the world through records in a database that are visually represented by points, nodes, and polygons on a map. With the abundance of spatial information available today, assembling two-dimensional and three-dimensional (3-D) maps that accurately represent our world becomes a trivial exercise. What is not trivial is the power that a GIS offers in being able to manipulate, ask questions, analyze, and represent information through this visual medium. Some problems concerning a project are readily apparent visually but not easy to interpret through columns and rows of numbers. Figure 1.1 depicts site topography using 3-D visualization.

REMOTE SENSING. Remote sensing is the science of measuring features of objects from a distance. This is typically accomplished through aerial photography and satellite imagery. Remote sensing devices can receive and transmit electromagnetic energy for scanning objects on the earth's surface. The advantages of using satellite imagery systems such as LANDSAT and the French SPOT satellite systems are that they can cover a large area and capture data in a short period of time. Because these systems can measure energy or radiation emitted from different features on the ground, they are an ideal source for capturing data such as vegetation, soil, and water.

One of the primary advantages of remote sensing systems, including aerial photography in GIS, is that data are already in digital format and can readily be used in different GIS applications. They provide high-quality stereo imagery

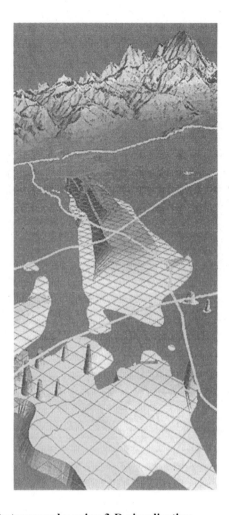

Figure 1.1   Site topography using 3-D visualization.

that can then be used to create 3-D models for use in developing topography and digital elevation models. These images can also provide coverage and data capture in areas where traditional access to the site may be difficult or impossible. Another advantage of remote sensing is its ability to ortho-rectify maps.

MODELING. Some of the first computers developed were used for modeling purposes long before GIS was invented. Because of this, there is a tremendous amount of modeling applications that have been developed that operate outside of most GIS. However, it has long been recognized that GIS can provide an abundance of information from which models can draw. In addition, GIS are perfect for visualization of modeling results. Many traditional hydraulic models that have been used within the water and wastewater industry are being integrated to take advantage of GIS. Although GIS technology is a logical data storehouse and user front-end to models, more important is GIS' ability to use modeling results with a variety of different data layers for enhanced decision-making. Integration of models within GIS is somewhat in its infancy. However, much of the data that a model requires are captured and stored as geospatial information within GIS. Therefore, in the future, models will be designed more seamlessly with GIS software as water and wastewater industry systems mature.

FACILITIES PLANNING. Planning for relief, expansions, or upgrades of existing facilities involves correlating an enormous amount of unrelated information to determine the facilities necessary to meet future demand. Using GIS technology as a tool for managing growth creates a more dynamic facilities planning process. Several spatial layers of information such as population growth, land use, and the location of proposed developments can be evaluated together to develop a single picture of anticipated growth within a service area. In addition, other topological features such as rivers, roads, ridgelines, and political boundaries can be used to delineate drainage and subdrainage boundaries. Geographic information systems technology provides tools for relating these individual layers of information together to determine when existing conveyance, distribution, or treatment facilities will require relief. The alignment and capacity of expansion to existing systems can also be determined using GIS through the use of hydraulic models, which integrate construction data, topography, and estimated flows also described spatially.

Geographic information systems are powerful tools for siting new public works infrastructure. Several layers of GIS data, constraints, can be correlated to evaluate the best possible location for new plants or pipelines. In this example, GIS is used to determine desirable locations by intersecting data layers to identify what areas are common to all. Information such as proximity to an existing conveyance system, surrounding land-use types, parcel information, sensitive habitats, utilities, availability of land, and land value are used to help narrow the possibilities. In this way, optimal sites for new facilities can be determined using many layers of unrelated information, with each layer containing a desirable quality.

PROPERTY MANAGEMENT. Most water and wastewater agencies are required to acquire and maintain a large amount of real property. Real property includes a collection of grant deeds, quitclaims, leases, and easements, which convey varied levels of rights necessary to construct, operate, and maintain water and wastewater conveyance, distribution, or treatment facilities. In addition, water and wastewater agencies may lease surplus property to public or private interests. These leases carry terms and limitations that govern the use of the property that must be managed. Geographic information systems technology allows for an intuitive map-driven application for managing property information. Through GIS, real property documents can be linked to property lines and polygons, which accurately depict real property boundaries. In addition, business transactions associated with leased properties can be tied in directly to property polygons. The GIS technology can also be used to maximize the value of surplus properties by performing a market analysis of the surrounding neighborhood, identifying potential markets within the community.

Property acquisition is a common use of GIS within property management. In this case, GIS can be used to evaluate potential properties based on comparable sales, property size, or other constraint criteria. The technology has also proven useful in acquiring land based on permitting requirements. As an example, a common permit restriction for operations within environmentally sensitive areas requires purchase and open space dedication of land of equal or better habitat value in exchange for land to be expanded upon. Habitat value can be spatially represented in a GIS and used to help identify desirable locations. Available parcels can be interacted with habitat data to identify potential sites that meet permit requirements. Environmental habitat data should be used, if available, from the permitting agency itself, which provides confidence in the selected parcels to be purchased.

EMERGENCY RESPONSE. After acts of terrorism in the United States in 2001, focus on emergency preparedness and response has shown that GIS are the tools of choice. Geographic information systems software was used extensively by emergency response crews in coordinating search-and-rescue operations. Later, GIS software helped manage logistics necessary for debris removal through map production and analysis (Harwood, 2001). The power of analyzing large amounts of data that describe our planet quickly and accurately makes GIS invaluable assets in a crisis situation. Geographic information systems can communicate the location, size, and severity of an emergency to decision-makers allowing them to direct resources in a timely manner. Geographic information systems are great tools for staging "what-if" scenarios to test emergency preparedness. In this case, GIS are used for assessment, communication, and response to a problem during a simulated emergency. Geographic information systems can also coordinate emergency-related information from various federal, state, and local government agencies. For example, federal and state disaster relief offices are developing Web-enabled services to deliver real-time disaster information in a spatial format. In the event of an emergency, an agency using GIS technology can spatially interact with disaster location and severity data with their own facilities to direct resources to where they are needed the most. In this way, the effects of a catastrophic event such as an earthquake or hurricane can be mapped out and integrated with an organization's infrastructure to assess damage in real time.

Geographic information systems technology has been used successfully in analyzing hazardous situations or the storing of hazardous chemicals within the work environment. Inventories of hazardous materials can be spatially rendered and used to assess risks to workers. In addition, the effects of a chemical release can be modeled within GIS to determine the potential effects to plant operators and the surrounding community.

SPILL CONTAINMENT. Water and wastewater utilities that have developed conveyance or distribution systems data can use GIS as tools for spill diversion or containment. The GIS software has the ability to discern direction of flow within a pipe network. Through this, GIS can accurately trace flow paths upstream or downstream from a predetermined location. One use of this technology is to build an application that can trace the path of a contaminant introduced to a conveyance system. In the case of separate sewer and stormdrain systems, GIS can isolate a contamination introduced to a stormdrain by a sewer overflow. Other uses of this technology include the ability to reroute water or sewer services. When it is necessary to take a portion of the collection system out of service, either for maintenance or repair, the same technology can determine the best method for redirecting flow through a pipe network.

SOURCE CONTROL. A significant challenge of operating a wastewater collection and treatment system is the ability to manage the discharge of industrial wastes. Geographic information systems technology has been used to protect these systems from illegal discharge of hazardous waste chemicals. Illegal discharges can be tracked down more efficiently using sewer-tracing functions within a pipe network in conjunction with a spatial database of permitted industrial discharges. Once a contaminant has been detected, GIS applications work to isolate the source of illegal dumping by identifying the sewers that convey wastewater to that point. Combining this information with profile data of chemicals associated with industrial process types will help narrow searches to a few candidates. Once identified, the same system can help narrow the search further by evaluating the collection system for upstream monitoring points that can be used to eliminate possible violators.

DEVELOPMENT MONITORING SYSTEMS. Geographic information systems technology has been used to extend the capabilities of a development monitoring system (DMS). Under current environmental permitting laws, water and wastewater agencies are required to determine the singular and cumulative effects of a proposed development on collection, treatment, or distribution facilities. A DMS program provides a tool for systematic cataloging of development information within GIS. Once cataloged, interacting with collection or service system boundaries can easily identify individual and cumulative effects to facilities. Pipe network traces are used to determine potential effects throughout a collection or distribution system by comparing available capacity with the additional flow.

Development monitoring system data within GIS also provide valuable insight to future demands of public works utilities. Large developments cataloged through the DMS process typically require two to five years to construct. Therefore, the DMS data provide a short-term planning horizon that is more tangible than other growth indicators. Progress of developments can be tracked within GIS using connection permit data and aerial photography. This type of application is especially useful in fast-growing regions where traditional sources of growth information are unreliable.

WHAT THE FUTURE HOLDS

Being able to anticipate the direction of GIS technology and its place in the water and wastewater industry is an important aspect of GIS development. In an industry that has reinvented itself every two to three years, keeping up with its current pace is a daunting task. However, use of GIS as assets in any public works endeavor is evolving toward a model of integrated data that is highly mobile. Today, computer systems within most water and wastewater agencies are working toward integrating information islands. Each island represents a core business function serving specific purposes within the organization. The GIS technology often acts as an adjunct to each of these islands by bringing information to the user through an intuitive interface. In the future, tapping into GIS' ability to communicate and relate an abundance of complex information across departmental boundaries will allow an organization to leverage its own information in ways that have yet to be imagined.

ADVANCES IN HANDHELD AND MOBILE GEOGRAPHIC INFORMATION SYSTEMS. For a long time, mobile GIS was one of those concepts that seemed just beyond the reach of most industry professionals. With the advances in hardware and software, it became increasingly evident that mobile GIS technology would soon become a reality. First came hardware devices such as rugged laptops and pen-based PCs. Next came the porting of major GIS software from mainframe and minicomputers, typically on a UNIX platform, to Windows NT platform. This was a tremendous step in making available sophisticated GIS software on portable computing devices, including laptops and pen-based PCs. Additionally, with the increasing accuracy and sophistication of global positioning system (GPS) devices, the necessity of integrating GPS with GIS and mobile devices became self-evident.

However, because of the physical nature of fieldwork, companies still needed devices that were more manageable and less expensive than the $3,000 to $6,000 price tag that laptops and pen-based PCs carried. It was the introduction and subsequent success of Palm handheld devices that captured everyone's imagination and, for the first time, sparked serious debate on the topic of mobile computing. These small and compact devices opened the door to a new era in mobile computing. However the Palm handheld device did not have sufficient processing power or the operating platform compatibility to allow for sophisticated computing such as GIS. With the introduction of Windows CE-based handheld pocket PCs and GIS software, mobile GIS took a giant leap in becoming a reality. With mobile GIS, databases become directly accessible to field-based personnel whenever and wherever information is needed. The GPS receiver and antenna are typically integrated to these Windows CE-based handheld devices, making them much more versatile and an ideal tool for field data collection. Windows CE-based handheld devices are being used with increasing frequency for diverse applications such as utility data collection for water and wastewater systems and road and pavement management systems. It has been shown that mobile GIS make economic and project sense to streamline the process of data collection.

Although hardware and software are getting more powerful in today's mobile field systems, wireless connectivity is still the weakest link in the mobile GIS equation. But that will change soon. Today's relatively limited bandwidth may be suitable for small jobs such as dispatch or simple work-order data management. However, they are not suitable for large file transfers such as CAD drawings and geospatial files. It is anticipated that this situation will change in the near future. As the public's appetite for mobile GIS and wireless connectivity increases, wireless vendors will unveil higher bandwidth "expressways".

With more than 40 million workers in the United States working in the field, developing the enterprise information infrastructure to allow for easy and seamless mobile computing will be a major objective in the next decade as well as one of the biggest opportunities to increase productivity, reduce costs, and add value to the enterprise (Wilson, 2002).

SEAMLESS DATA INTEGRATION TECHNIQUES. The phase "seamless data integration" seems to be an oxymoron. Lack of uniform platforms and standards make it difficult to move data to and from different media. It was not so long ago that complicated and expensive translators were needed to translate CAD data to different GIS software and relational databases, and GIS software more often than not seemed to be at war instead of coexisting peacefully. However, with the increasing sophistication and flexibility of both GIS and CAD software, it is now a lot easier to move between the two media. Major GIS packages can read CAD data directly and there are third-party applications that allow for CAD-like manipulation and design within GIS. Also, GIS and major relational databases can communicate and pass data seamlessly, thereby combining the best elements of both.

Data integration has also affected the mobile and handheld computing arena. With better and more powerful software, handhelds and desktops are communicating effortlessly. Using Windows CE-based operating systems, applications such as Visual Windows CE and Windows CE-based GIS are making field data collection a painless enterprise that is increasing efficiency and improving productivity at a rapid pace. The next evolution in the GIS data integration arena will be improved and inexpensive wireless technology. This will allow for enhanced communication and update in real time and will make the jobs of operations, maintenance, and customer services of public organizations more effective and productive.

Although still far from reality, it is getting a bit easier to bring different data sources together. Still, this is an area for future development and innovation and an area in which the hard work and innovative solutions will be needed to achieve the ultimate goal of "seamless data integration".

GEOGRAPHIC INFORMATION SYSTEMS PROLIFERATION IN THE WORKPLACE. People throughout the world are realizing that they can use

GIS as tools to make their everyday tasks more efficient. Geographic information systems are becoming commonplace in surveying, planning, public works, and engineering. Some environmental applications include modeling of coastal erosion and restorations and modeling of dredging operations. Additionally, fire and police departments are using GIS for modeling wildfires and crime analysis, respectively. Powerful applications are being developed in health sciences that take advantage of GIS spatial and analytical capabilities. These include applications for research and control of malaria and a cancer geographic information system. This has led to a surge in interest and the pursuit of the science of GIS. There has been an increase in the number of GIS courses being offered by various departments in different colleges and universities. These educational institutions have also tried to address the growing demand for GIS knowledge by working professionals through establishing GIS certificate programs. The end result of these activities has been the proliferation of GIS software in the workplace. People are increasingly relying on GIS software to perform their daily routines, such as

- Tax map management and tax assessment activities;
- Customer records and service;
- Planning, zoning, and property management;
- Utility modeling and mapping;
- Environmental planning;
- Asset and infrastructure management; and
- Infrastructure system rehabilitation and design.

This trend will continue as people and departments become more connected through enterprisewide GIS and find more uses for developing and analyzing spatial data.

ENTERPRISE GEOGRAPHIC INFORMATION SYSTEMS SOLUTIONS. To operate effectively, information must be viewed as a resource and shared by all departments and divisions within an organization to support many different functions, while keeping pace with public demands. Future success will be defined by one's ability to leverage the core knowledge base, while understanding how organizations receive and assimilate information. Viewing information as a resource requires a change in the mindset that is prevalent through all levels of government regarding data. This includes changing our perceptions by forcing us to think about work functions and priorities from an enterprisewide perspective. Enterprisewide implies that organizations act as a whole, moving with purpose in a defined direction. An example of an enterprisewide goal would be land-use planning among all governments that have an interest in an area. Another example would be sharing water system inventory and attributes between engineering and fire departments.

According to many GIS insiders, the GIS industry is moving from data sharing at individual levels to data sharing at enterprise or organizational levels. The visions of how and what users share varies, but a few common threads emerge:

- The standard database now plays a leading role in data storage and management. Individual CAD files and formats will play a more limited role.
- The database will become the "campfire" around which spatial data users sit—with client software on laptops and handhelds.
- Client software likely will look different from CAD, although it may share a few of its "best parts".
- GIS is moving towards "GIS": As GIS software is integrated with traditional information management systems, geography becomes less important.

With increasingly limited funds and resources, agencies are asked to do more with a lot less. That is why the trend towards an enterprisewide GIS has started and will continue in the foreseeable future. It makes logical as well as economical sense.

SOCIETAL GEOGRAPHIC INFORMATION SYSTEMS. Societal GIS is seamless integration of GIS resources across local, county, state, and federal government. It requires accurate metadata, infrastructure, established clearing-houses, and changes in attitudes with regard to sharing data.

Virtually everything we do is inherently tied to a location and it is this spatial component that will be the key to combining data from many sources. There are many examples of societal GIS at work today. For example, many people can use MapQuest to find the address and route between two locations. That is a good use of a societal GIS application. On the flip side, marketing and advertising agencies are using those neatly collected cookies you leave behind while surfing the Internet and developing a GIS-based marketing application for a targeted marketing blitzkrieg customized to your taste. Within a short timeframe, a real-estate sales agent will be able to call up a property's registered survey, show a photograph of the property (perhaps even take a virtual tour), find the nearest school or school bus route, show the city's official plan zoning for the neighborhood, find all properties that meet a purchaser's price and design criteria, and advise all affected utilities if a change of ownership transpires. This will not require any more knowledge and training on the real-estate agent's part than learning how to use a modern word-processing program. Societal GIS will no longer be a tool for the mapping professional; rather, it will be a product or a service that anyone can use.

The appeal of societal GIS is in its widely available access, which is free of charge. You will pay only for the services that you use. There are many analogies in our present-day culture. The prices for many consumer electronics such as PCs, gaming devices, and cell phones are plummeting. However, you pay for the software in the case of PCs and game cartridges for gaming hardware and the airtime and network charges in the case of cell phones.

ASSET MANAGEMENT. Asset management is a business process and a decision-making framework that covers an extended time horizon, draws from economics as well as engineering, and considers a broad range of assets. The asset management approach incorporates the economic assessment of trade-offs among alternative investment options and uses this information to help make cost-effective investment decisions. Asset management is usually described as the overall long-term vision, operating philosophy, and general direction to be used in managing an asset. State and local governments spend $140 to $150 billion a year in construction, improvement, and rehabilitation of the public's capital assets, including

- Sanitary sewer systems,
- Water systems,
- Storm sewer systems,
- Traffic control devices (signs, traffic signals, etc.), and
- Roads.

Therefore, asset management can have real strategic benefits for government organizations. These benefits include improving information access and content, focusing strategic initiatives, and helping develop faster decision-making. Additionally, strategic asset management can help identify economies of scale and possible alternative courses of action, guided outsourcing and privatization activities, and help identify the best potential to expand services offered.

For an asset-management program to be effective, it needs to have the following components:

- Developing and tracking of an inventory of desired assets,
- Maintaining available information on the condition of assets,
- Determining the value of assets,
- Effective use and analysis of collected data,
- Properly analyzing the cost of maintaining an asset over the cycle of its life, and
- Analyzing costs and benefits of a project and determining the best course of action.

Ultimately asset management helps develop flatter and faster decision-making. It also forms the basis for systems perspectives throughout an organization by integrating departmental activities toward clear organizational objectives.

DYNAMIC FACILITIES PLANNING PROCESS. Ideally, when using GIS as a tool for facilities planning, the analysis should rely on information from multiple jurisdictions. Coordinating information among different local, county, and state agencies requires the adoption of rules and standards that must be followed. However, the benefit of societal GIS is in its ability to provide up-to-date data from many levels of government that can be related together seamlessly. Information such as population growth estimates, land use, and the location of proposed developments are good sources of information that help determine when, where, and to what capacity new facilities will have to be constructed. Theoretically, a living model of a facilities plan can be constructed and maintained by establishing the necessary files that are maintained within a societal GIS. As information is updated within this system, facilities planning can be updated automatically to reflect the changes that were made. In this way, facilities planning can be updated with the most relevant information from jurisdictions that are responsible for their respective pieces. For example, land use within a service area is created through local general plans. Typically, planned upgrades to public works facilities rely on general plan information that describes an ultimate flow condition. However, general plans are not static but change over time to reflect the changing needs of the areas they describe. Because of this, it is important that changes to general plans are also reflected in the plans that utilities have for future service to the jurisdictions they serve. By using GIS technology, the facilities planning process can be responsive to the changes within a community that it serves and rely on the most relevant up-to-date information to plan for future needs.

EMERGENCY RESPONSE. During the aftermath of a shooting at Columbine High School in Littleton, Colorado, emergency response teams determined the single most valuable resource that was not available but could have made a tremendous difference in the outcome (Feliciano, 2002). This resource was the availability of up-to-date maps that depict the school's layout. In response to this, emergency teams now carry a wide array of maps that will help them in future incidents. With advances in mobile GIS technology, the availability of map-based information will only get better. Imagine a societal GIS that shares spatial data from numerous public agencies through a universal map application. In this scenario, emergency responders could have tapped into the schools AM/FM application that contained up-to-date facilities information, including the layout of the school, number of students, and perhaps a schedule of classes. Recent advances in imagery technology could have provided photographs of the school shot at oblique angles from just about every direction imaginable and with a resolution so fine that you would be able to spot a golf ball lying on the school grounds. This photo imagery would also give the user the ability to measure walls, windows, and rooftops. This application would go on to help deploy law enforcement and rescue personnel at key locations throughout the school. This is not some far-fetched futuristic vision, but technology and applications that exist today. The same lessons learned through the Columbine High School incident apply to any emergency situation that water and wastewater utilities may face in the future. The technology exists today to help provide the maximum amount of relevant information, which can save lives.

TOTAL MAXIMUM DAILY LOADS. Most water and wastewater utilities that operate treatment facilities are facing a new framework of discharge regulations known as total maximum daily loads (TMDLs). Originally part of the Clean Water Act, TMDLs approach water pollution from a mass-balance perspective. The TMDL process examines a watershed from a macroscopic level and is triggered if the U.S. Environmental Protection Agency (U.S. EPA) has determined that any water of the United States has reached a level of impairment. Once triggered, the TMDL process seeks to manage discharges of the chemical constituent identified as the cause of impairment by imposing further restrictions, if necessary, to bring the water body into compliance. U.S. EPA has used GIS as a tool for modeling both point and nonpoint source pollutants within a watershed to facilitate the TMDL process. This program, known as Better Assessment Science Integrating Point and Non-Point Sources (BASINS) is a GIS application that ties water-quality data, the National Pollutant Discharge Elimination System discharge database, the Toxic Release Inventory, and Superfund sites together with base map information for all major watersheds of the United States. The BASINS program can be used to analyze the information contained to help determine further restrictions, if necessary, on discharges that will result in the improvement of water quality.

RESOURCE MANAGEMENT. Efficient management of water resources is a key element for a water utility's ability to provide a reliable drinking water source for customers. For most water utilities, water resources come from a variety of places, some of which may vary in the amount of water yielded on an annual basis. Because of this, it is necessary to be able to anticipate short-falls and surpluses of available water resources. Geographic information systems technology provides an efficient tool for balancing supply with demand for any water utility. By establishing need within a service area, GIS applications can help anticipate how water shortages can affect a community. In addition, GIS can help ensure that an adequate supply of water is available to meet the future needs of the community by integrating resource management with facilities planning information. Allocating a water resource based on water rights could be factored into this model, which can help navigate legal issues, especially complex in the western United States.

OBJECT-ORIENTED GEOGRAPHIC INFORMATION SYSTEMS. Being able to tap into GIS technology through object-oriented programming is relatively new. Object-oriented programming refers to specific parts of any program, such as spell checking within a word processor, which are segregated into groups of tools known as objects. These tools can be related to each other through a simple set of rules or commands. In this way, applications are far more customizable (one can assemble objects to make an application).

Object-oriented GIS provides added functionality to both database modeling and programming. New behavior objects allow for better database modeling by adding additional variables that mimic real life. For instance, the geodatabase model allows for rule-based information to aid in developing spatial information. An example of this would be to enable GIS software to know that a 200-mm (8-in.) connection could not be located on a 100-mm (4-in.) pipe. In general, programming itself is enhanced greatly by being able to take advantage of GIS objects that allow its technology to be integrated as a part of a specific application. An example of this is how an object-oriented GIS can integrate with outside objects to extend the functionality of an application. For instance, model integration could be facilitated much easier if model functions were encapsulated within an object. This object could then be easily integrated with GIS objects to form an application. The result is software application in which the modeling functions would appear to be originally built within the application itself.

MODEL INTEGRATION. There are many examples of hydraulic models being integrated with GIS technology. Applications have been developed that work outside of GIS software but draw upon information stored within GIS. This is a logical approach as most hydraulic models seek to describe a spatial circumstance. Such applications have taken this one step further by either manipulating GIS data in real time or creating an output that is accessible through GIS software. Recognizing that most water and wastewater agencies are migrating toward geospatial systems, hydraulic modeling software manufacturers have begun to develop software suites that work more seamlessly within GIS. These packages can be loaded within GIS, extending its capabilities to include hydraulic models. However, any model is only as good as the data that it is drawing upon to reach its conclusions. Care should be taken when deciding on the level and accuracy of information that is a part of any GIS if hydraulic modeling is to be used.

To truly understand the hydraulics of a sewer system, a fully dynamic model is required. This will help plan and define improvements to the system to accommodate future growth. Physical attributes such as pipe size, length, invert elevations, and materials are contained within GIS. The GIS models can be integrated with supervisory control and data acquisition systems, which deliver live streaming data for analysis, visualization, and model calibration. Flow-monitoring sites can be accessed and correlated to upstream basin information. Finally, interpretation of results can be graphically displayed to identify upgrade needs.

SECURITY ISSUES. Terrorism within the United States is a concern that has been heightened. Water supply as well as water and wastewater treatment facilities have been identified as potential targets, which could affect a large population. Many agencies have since taken measures to increase security at plant facilities and water storage and distribution centers. In the past, GIS technology has been deployed to help in security situations. During the 2000 presidential campaign, GIS technology was used successfully by the City of Los Angeles, California, in managing demonstrations outside of the Staples Center during the Democratic National Convention (Wolfe and Mitchell, 2000). It was a valuable asset for crowd control and diversion as well as deployment of security personnel.

Geographic information systems can also analyze existing facilities to probe for weaknesses or lapses in security measures being taken. Analyzing strategic locations around water and wastewater facilities can be determined through plant access information, terrain modeling, and the siting of critical equipment. Different scenarios of a terrorist act can be modeled within a GIS, including the presence of biotoxic agents introduced to water storage or distribution systems.

The future of GIS as a deterrent for terrorism could be hindered by the need to departmentalize sensitive information. Since September 11, 2001, the trend of making geographic data widely available has reversed for fear that the very data that would allow us to probe for weaknesses could also be used against us. For instance, detailed information on the layout of a water treatment facility could help terrorists identify a location to introduce a biotoxin to the water supply.

Sensitive data can be shared if proper precautions are followed. For instance, the Federal Geographic Data Committee suggests that data on the nation's infrastructure can be shared responsibly and provides a tremendous value for deterrence. Examples include GIS applications that facilitate detection of potential targets, offering tools for emergency planning and preparedness, identifying vulnerable situations to prevent an attack, and improving response times through mock situations.

*R*EFERENCES

Dodge, M.; Doyle, S.; Haklay, M.; Rana, S. GIS Timeline. http://www. gistimeline.org (accessed 2003).

Feliciano, B., Crime Mapping Division, Rio Hondo Community College (2002) Personal communication.

Harwood, S. (2001) New York City—Creating a Disaster Management GIS on the Fly. *ArcNews*, **Winter 2001/2002**.

Wilson, J. D. (2002) Industry Innovations, Mobile Technology Takes GIS to the Field. *Geoworld*.

Wolfe, M.; Mitchell, T. (2000) Applications for the Democratic National Convention; Los Angeles County Environmental Systems Research Institute User Group Seminar.

Chapter 2
The Value Proposition of Geographic Information Systems

FUNDAMENTAL PURPOSES SERVED BY GEOGRAPHIC INFORMATION SYSTEMS

Use of geographic information systems (GIS) serves at least nine fundamental purposes that provide value to users. These fundamental purposes include mapping and databases, facilities management, facilities atlases, management decision-making, facilities planning, federal regulation compliance, business process reengineering, public perception, and e-commerce.

THE VALUE OF GEOGRAPHIC INFORMATION SYSTEMS FOR MAPPING AND DATABASES. Geographic information systems, which can store any type of information with geography as its common denominator, have two primary functions that provide value

- A GIS functions as a central repository for maps and databases. At the most fundamental level, GIS provide value because they store geographic information in graphic (maps) and nongraphic formats (tabular data) in one place. In addition, GIS allow maps and databases to be linked or integrated. This integration allows for the creation of intelligent maps that can be queried and analyzed at a much deeper level than paper maps, computer-aided design (CAD) maps, or tabular data alone. Also, once a central repository has been established, it represents a knowledge base that preserves and enhances information that has been stored over the years.

Geographic information systems also have value because they have all of the power of a database, which allows automatic data access, manipulation, query, display, and output. Inside GIS, disparate datasets of varying sources and types can be integrated, displayed, accessed, analyzed, and evaluated in meaningful ways.

- A GIS functions as a portal to many different kinds of intelligently connected databases. Geographic information systems also function as views or windows to additional data, such as financial records, stored in databases outside GIS that do not have a geographic component. The GIS tools allow users to view and display these data in conjunction with spatially related data in GIS so that trends can be spotted and conclusions drawn.

Geographic information systems offer a virtually unlimited array of display and output options because they distinguish between how data are stored, displayed, and output. Inside GIS, data types are grouped and stored separately from other data types. Unlike CAD data that must be displayed and output exactly as it is stored, with specific line weights, colors, and font sizes, GIS can display and output data by any attribute information stored in the database using any user-defined symbol or font. Thus, a water main can be color-coded by diameter or material with a simple point and click operation—there is no need to create a separate layer for each.

By selecting only certain data layers within GIS, thematic maps can be created for specific applications. Thus, users can create different kinds of maps using the same source data. For example, users can create a $1'' = 200'$ scale map from the same data used to create a $1'' = 1000'$ scale map. However, GIS would be used to vary the amount and type of information mapped at the different scales because it is feasible to show more detail on the $1'' = 200'$ scale map than on the $1'' = 1000'$ scale map. Attribute assignments can be varied to color-code data as desired.

Geographic information systems are structured, centralized repositories that, with careful and planned upkeep, can provide an accurate, up-to-date information inventory. During design of the GIS database, a GIS developer will use business rules and relationships to define how data must be input now and maintained in the future. As geographic features or areas within the GIS change, they can be edited and updated as needed.

Geographic information systems maps and databases can become even more useful over time if additional information—such as parcels, land use, zoning data, population data, and topography—are added to enhance existing GIS maps and databases that might contain water, sewer, storm, and other utility infrastructure information. In addition, GIS resources are easily deployed out in the field through intuitive user-friendly applications.

THE VALUE OF GEOGRAPHIC INFORMATION SYSTEMS FOR FACILITIES MANAGEMENT. An organization that wants to use GIS to better manage its facilities must first know what facilities it has so it knows what facilities it needs to maintain. Geographic information systems provide value by allowing users to maintain a graphic and nongraphic inventory of its facilities and receive, process, and track workorders for repair and maintenance activities. For example, if a water main break has occurred at a particular intersection, GIS allow a user to spatially locate the facility, find the best route to the facility, and use spatial tools to review the area and adjacent facilities to determine what work tasks need to be performed and what vehicles and equipment must be taken to the site. Instant access to GIS information allows utility managers and field crews to plan maintenance activities more efficiently and make better decisions both in the office and in the field. In addition, pipe traces can be automated within GIS, allowing crews to isolate breaks by rerouting flows.

The GIS maps can also be downloaded onto pen-based personal computers (PCs) for reference by field crews. For example, GIS data can be reviewed to determine conditions, other structures affected by the repair or maintenance activity, and historical records showing past repair and maintenance tasks. Field crews can also use pen-based PCs to record information about the current repair or maintenance task so that data can be downloaded back to the GIS. By having instant access to accurate, updated records, crews can do their jobs more efficiently.

THE VALUE OF GEOGRAPHIC INFORMATION SYSTEMS FOR FACILITY ATLASES. Plant site owners increasingly have critical needs for mapping selected site features. The preferred method of addressing this need is through creation of digital representations of buried utilities and selected features. These digital representations are referred to as *facility atlases*. The need for GIS-based facility atlases is perhaps less considered but is equally important as that of building GIS water distribution and wastewater collection systems networks. For example, facility atlas piping is typically not contained within street right-of-ways; there is often a greater number of piping systems and a greater need for quick access to piping information to support facility operations and maintenance.

Without a facility atlas, facility owners typically use as-built drawings as the record of buried utilities and related subsurface site features, and this often leads to problems

- Each facility site has often been built by multiple projects, so there is no single set of up-to-date as-built drawings to represent buried utilities.

- Without this single source of up-to-date data, finding an answer to a simple question is often a time-consuming task, typically requiring research through a number of paper drawings and CAD files.
- In the absence of documentation of as-built conditions, field verification, including potholing, is required to support projects or maintenance activities. Field verification of the many buried utility systems and process interconnections can require significant resources.
- Failure to find information or accurately determine as-built conditions can result in costly problems during facility operation and maintenance as well as design and construction of facility improvements.

Facility atlases offer the ability to quickly locate buried utilities and related data and drawings. It is important for facility owners to be able to quickly locate information about assets in the field and, in particular, buried utilities and structures. Facility atlases that spatially locate assets and provide user interfaces to data and drawings significantly aid owners in daily activities in determining locations, capacities, and costs. Typical uses for a facility atlas include

- Identifying utilities and property easements at a location within the plant site in advance of digging.
- Planning and implementing isolations of piping systems in support of maintenance and construction activities.
- Planning and designing alignments for new utility piping or locations for new structures, based on the facility atlas locations of existing piping and structures.
- Planning and designing connections to existing piping systems in support of facility improvements.
- Providing a database for recording maintenance information regarding the assets and performing analyses (e.g., financial analysis and asset management).

THE VALUE OF GEOGRAPHIC INFORMATION SYSTEMS AS MANAGEMENT DECISION-MAKING TOOLS. Geographic information systems provide great value to managers responsible for tracking and managing infrastructure assets. By reviewing accurate and complete GIS data, such as the inventory, condition, and value of assets owned, managers can make better decisions.

Geographic information systems can be used to automatically track asset values as they change and are updated over time.

Geographic information systems can be programmed to answer management's frequently asked questions and one-of-a-kind queries. For example, the query "show me all of the sewer pipes installed last year" would produce a map indicating the locations of all the pipes installed last year so management can spot trends at a glance. Other queries might be "show me all former utility assets that have been abandoned", "show me all utility assets that have been replaced in the last five years", "show me all closed valves throughout this service area", "show me all customers that would be affected by a shutoff of valve 12345", or "show me all properties that fall within X watershed". The complexity of the question is limited only by the availability of data.

By querying the GIS, management can generate specific maps required for decision-making or projecting the effect certain decisions will have on existing infrastructure. By performing a spatial or tabular query and displaying the output on a map, results suddenly become more meaningful and relevant than when displayed in columns and rows. Because accurate and complete GIS data are being used as the bases for decision-making, managers will likely have more confidence in their decisions.

THE VALUE OF GEOGRAPHIC INFORMATION SYSTEMS FOR FACILITIES PLANNING. Geographic information systems can be used to plan the repair and replacement, extension, or addition of new facilities resulting from population growth and new development.

As a facilities planning tool, GIS can be used to analyze historical information collected for facilities–maintenance management applications and tracking information collected for asset management applications. In the past, facilities planning was primarily a manual process often based on instinct, hunches, and some historical knowledge of the system. By using GIS tools to perform spatially based queries, users can determine where improvements are needed based on projected population, general plan information, proposed developments, or maintenance histories. Geographic information systems offer great value when performing utility master planning and capital improvement planning by combining data collected in the past with future projections.

Geographic information systems are most valuable for facilities planning when analyzing several datasets at once. Thus, a GIS allows decisions to be made using several factors (data layers) that more realistically model real life. Once managers have drafted a short- or long-term capital improvement plan, recommendations can be mapped so that constituents can see graphically the effect of the plan on the service area.

Hydraulic modeling is often performed as part of the facilities planning process. Geographic information systems can add value to the modeling process if accurate and complete GIS data are used to build and feed the modeling network. Using GIS data to build the network produces more refined modeling results, which can then be exported back for display and output.

THE VALUE OF GEOGRAPHIC INFORMATION SYSTEMS FOR COMPLYING WITH FEDERAL REGULATIONS. There are several new federal regulations with which GIS can assist with compliance. These federal regulations may require additional work to be performed by the utilities, but they can also be regarded as potential funding sources to help build or enhance GIS. Only brief overviews of two recent regulations, CMOM and GASB 34, are provided here.

Governmental Accounting Standards Board Statement Number 34.
Governmental Accounting Standards Board (GASB) is a private, nonprofit organization formed in 1984. It develops and improves accounting and financial reporting standards for state and local governments. Although GASB is not a federal entity, governments must follow GASB standards to obtain approved financials from their auditors. In June 1999, GASB issued its statement number 34, which represents the most comprehensive and far-reaching accounting rules ever developed for governments. Statement number 34 requires state and local governments to establish a mechanism that reports infrastructure value, depreciation, and management practices. The most important new rules involve reporting of general infrastructure assets. Among its many new provisions, GASB 34 will now require governments to report the value of infrastructure assets such as water and sewer facilities, roads and pavement, and similar long-lived assets.

Key features of GASB 34 requirements include

- Inclusion of infrastructure (value) in the asset base report in annual financial statements,
- Reporting of infrastructure assets at historical cost or estimated historical cost,
- Determination of significant general infrastructure assets at the network or subsystem level, and
- Following initial capitalization, depreciation or reporting of infrastructure assets using a modified approach.

The GIS processes that can be used to assist with this process include asset inventory and valuation models, depreciation, and implementation of an asset-management system in lieu of depreciation.

Statement number 34 requires that all capital assets (including infrastructure) be capitalized at historical cost (or estimated cost) and depreciated over their lives in use. An alternative to this depreciation method is the use of the modified approach. The modified approach requires the government to demonstrate that it is maintaining the infrastructure at or above a condition level that has been established.

The modified approach places a great deal of emphasis on on-going maintenance of the assets. Deferred maintenance has been shown to be more expensive than ongoing maintenance and renewal and rate of deterioration increases with time. That is why the modified approach or asset preservation method is preferred by utility departments, public works, and engineering. In short,

- Continual maintenance delays deterioration and failure,
- High replacement costs are delayed,
- Failure costs are avoided (fines, public relations, damages),
- Expensive emergency repairs and replacement are avoided, and
- Emphasis is on maintenance, hence extending the life of assets.

Capacity, Management, Operation and Maintenance. Capacity, Management, Operation and Maintenance (CMOM) of wastewater systems is a term used in U.S. Environmental Protection Agency (U.S. EPA) regulations for sewer system overflows and an amendment to the National Pollutant Discharge Elimination System (NPDES). Whereas GASB 34 is not a federally mandated program, CMOM, once approved by U.S. EPA, will be a federal mandate. The objective of CMOM is to improve utility business practices. In summary, CMOM requires the following:

- Project information summary,
- Utility description,
- Utility profile,
- Recent performance summary,
- Management programs,
- Operational programs, and
- Maintenance programs.

Both GASB 34 and CMOM require inventory and management of assets, an assessment of condition, procedure for identifying maintenance and repair schedules, and investment plans. Both of these programs are data intensive. Because islands of data exist in many different departments and locations such as CAD, supervisory control and data acquisition (SCADA), computerized maintenance management systems (CMMS), finance, and engineering, there needs to be a great deal of coordination and cooperation within organizations to make these efforts successful. Increasingly, agencies are turning to GIS as the central element for implementing GASB 34 and CMOM programs. Using GIS, agencies can record and track the needed data inventory and georeferencing of their assets. Furthermore, they can perform a comprehensive data population of asset attribute such as condition, age, date, and maintenance history and schedule. Ultimately, enterprisewide GIS can enable data sharing and data updates across an organization's network, eliminating the need for duplicate databases as well as improving the organization's overall effectiveness and performance.

THE VALUE OF GEOGRAPHIC INFORMATION SYSTEMS FOR BUSINESS PROCESS REENGINEERING. An existing business process that is inefficient will not instantly be improved by using GIS. Developing GIS should be considered an opportunity to investigate and reevaluate existing processes and workflows built on legacy systems, determine the real value steps, and use GIS tools to create a more efficient workflow through automation. The real value of GIS comes when a time-consuming 12-step process can be reduced to a five- or six-step process, saving both time and money.

Typical processes that are often reengineered for GIS include maintenance and workorder processing, permitting and plan approvals, and construction inspections. While GIS developers can plan how an existing process might be reengineered if GIS is applied, the new process cannot be implemented until GIS is in place.

THE VALUE OF GEOGRAPHIC INFORMATION SYSTEMS FOR ENHANCING PUBLIC PERCEPTION. Most of the general public is still unaware of what GIS is, and thus declaration by a government that GIS is in use will probably not enhance public perception. However, improved efficiencies and conveniences experienced by the general public by using GIS will definitely improve the perception of an agency. So it is not as important that the public know that it is GIS that is improving their customer satisfaction as it is that the public acknowledges improved customer satisfaction and the agency recognizes that it is because of GIS.

The GIS tools can be used to create credible, professional, custom maps—based on accurate and complete data. The media and the general public will find technical plans and proposals easier to understand if GIS data are used to produce tailored maps that depict current problems and the positive effect of proposed improvements. For example, by depicting graphically and spatially where frequent repair and maintenance activities occur, the extent and location of flooding zones or neighborhoods with water quality problems, an organization can more easily explain and justify costs for capital improvements.

The public will have more confidence in an organization's maps and records—and perceive an organization as providing outstanding customer service—if staff can fulfill requests instantly by accessing GIS. The public will be even more pleased with the organization if GIS maps are provided on demand through the Internet or conveniently located public-access terminals. If developers, realtors, plumbers, construction companies, or the general public must stand in long lines for information, travel from office to office to access maps, or wait on clerks to sort through paper maps in drawers, they will perceive the organization as inefficient. Conversely, an organization that has GIS data at its fingertips will be perceived as a modern, progressive, efficient organization worthy of continued public support. What is more, communities that have easy, electronic access to maps and records through GIS will likely be more competitive and successful in attracting new business and economic development.

THE VALUE OF GEOGRAPHIC INFORMATION SYSTEMS FOR E-COMMERCE. Geographic information systems technology can add valuable information to any e-business solution. Data from GIS can be bundled and shared with the public easier than paper maps. For a nominal charge that varies from state to state, developers and other customers can review available GIS data over the Internet, determine what GIS elements they need, and make an online request for a CD-ROM containing a portion of the GIS for a particular application. This eliminates the need for users to spend time traveling to several offices or departments to collect the mapping information they require. In addition, map-driven applications can add an intuitive interface for e-commerce. This can lead to quicker user acceptance and increased frequency of use.

BENEFITS DERIVED

The GIS allows one to manipulate and view information in a graphical format to produce a highly detailed map. Moreover the end user or analyst can directly manipulate the map before contracting with a cartographer or graphic artist to finalize the publishing of a map. A highly skilled technician can maintain data integrity and accuracy. But GIS is much more than just maps. First and foremost, GIS can support accurate, reliable and shareable information. This information can be graphical data or it can consist of nongraphic "attribute" data. These attributes provide additional descriptive information about the nodes, arcs, and polygons depicted on the map.

The GIS can provide us with useful information about the world around us. If a picture is worth a thousand words then, by extrapolation, a GIS picture must be worth at least a million. To demonstrate trends and patterns in a visual context, we can convey information to people much more effectively than just a table of numbers and words.

Probably the most important benefit is that GIS puts accurate, reliable, useful information in the hands and on the desktops of the people who can use it and need it the most.

Geographic information systems brings together the entire organizational information database from record plans in the map file room to a customer letter in an organized and concise format delivered to everyone's desktop.

Geographic information systems can support

- Facilities management,
- Workorder management,
- Logistics management,
- Environmental management,
- Condition assessment,
- Utility modeling,
- Corridor analysis, and
- Monitoring and compliance.

Geographic information systems can help to

- Improve operating efficiencies,
- Provide improved customer service,
- Eliminate redundancies,
- Consolidate information,
- Integrate applications, and
- Assist in problem solving.

The specific benefits of GIS can range widely depending on the initial targets and objectives set forth in the implementation and planning stages of development. For the most part, benefits of any project can be categorized for better understanding. A benefit can be either quantifiable or nonquantifiable and tangible or intangible. For example, we can measure items such as cost reductions, operating costs, staff time, and increased revenues. The intangible benefits, which are hard to quantify, typically are associated with improved decision-making, decreasing uncertainty, or effect on the corporate or organizational image.

We can speak of benefits being direct or indirect. A productivity improvement to a department would be a direct benefit to that department and indirect to the overall organization or customer group.

In this section, we are examining the specifics of benefit derived from using GIS. The topics are outlined as follows:

- Movement of the industry to a business model,
- Centralized data resources,
- Application integration,
- Streamlining existing workflow,
- Improved customer service (workorder management, emergency response),
- Maintenance management systems (tracking complaints, mapping resource),
- Closed-circuit television (CCTV) scheduling and information base,
- Cost management (budgets, condition assessment),
- Plan approval and permitting,
- Mobile computing,
- Project visualization,
- Coordinated emergency response, and
- Enhanced decision-making.

Each one of the subsections is a derived benefit from implemented and functioning GIS. The subsections will expand on the topic to identify and outline some of the specifics and site examples.

BUSINESS MODEL. The wastewater utility has been migrating to operation as a self-sufficient utility or business model over the past several years. The concept of full cost recovery and asset management are coming to the forefront as managers struggle with financial commitments of a constantly degraded infrastructure and charge the customer a "fair" market price for the utilization. So where does GIS fit in this picture? Quite simply, GIS becomes management tools to track the utility's assets, customers, and operation and maintenance initiatives. As a physical asset-management tool, GIS can be developed to detail the characteristics of the collection system and associated facilities. Geographic information systems can begin as simple mapping tools and migrate to full data inventory systems. All possible attributes can be associated to GIS, including as-built information on pipe materials, lengths between manholes, manhole locations, invert elevation, and rim elevations. We can start to track maintenance activity to specific reaches of the collection system creating a historical database. This may lead to improving or replacing this section of pipe, thus reducing maintenance costs for this particular reach. We have not only addressed the issue of maintenance spending but have associated this to a physical entity in our system.

The reality of the situation faced by wastewater managers is that they need to "do more for less". Being able to target specifics is one way of reaching this objective. Examine the way your line of business operates, prioritize the work stream, eliminate non-essential functions, and improve on the delivery of essentials. Acquisition, storage, and analysis of spatially referenced material are essential in a utility. Manual methods have their limitations such as limited data evaluation, redundant data sets, lack of consistency, map degradation, data from various sources, and different scales used in coordinate systems and aerial coverage. Because of these factors, management and sharing of essential information can be difficult and expensive. Effectively implemented, GIS can address these issues and improve the overall workflow. This all comes at a cost that, of course, needs to be carefully weighed against the benefits.

CENTRALIZED DATA RESOURCES. The advent of network computer systems, whether they be local area networks (LAN), wide area networks (WAN), or Internet-based has lent itself to centralize data storage and retrieval. This does not mean that all data have to reside in one location. The resource will be reliable and timely. We have come full circle in the computing world, beginning with mainframe systems then moving this computing power to the desktop only to realize that the true "power" is to share and communicate data that we create. As the means to facilitate this "network" have become a reality, we can now share information and data with ease. This also makes logical sense from the perspective of using base data that do not change and only concentrating our personal efforts on the value-added component. This hierarchy of security and data version management, with each person assigned responsibility for their own data, will lend itself to further efficiencies in generations of maps for any purpose. No longer will you wonder how accurate the legal coverage or theme is because its maintenance is another's responsibility and function.

Now that data are centralized, they can be managed more efficiently, kept current, improved in accuracy, and eliminate duplicates. The documentation of this process resides in the "metadata" file inherently attached to the base data. Metadata facilitates the transfer of information by communicating what is available, its accuracy, and who maintains it.

APPLICATION INTEGRATION. A wastewater utility will use several computer applications to streamline the workprocess and enable its employees to "work smarter". Management is always examining ways to more effectively manage day-to-day and long-range operations. By virtue of the business, each department becomes responsible for its own specific line of business and will implement cost- and time-saving measures. Such applications can include facility management, SCADA laboratories, customer information, sewer inspections, and CAD system. The bottom line is to provide better and more efficient service to customers. The potential problem with these systems as a whole is that they are typically independent of one another. Because the majority of data is inherently spatial, GIS can help play a role in integrating all of these applications to one central access point.

For years, utilities have used billing databases to keep track of customers and their accounts, but have you ever wondered where customers are in the collection system? Geographic information systems can provide this answer with the integration of billing system information. This will also lend itself to proactive customer service if performing maintenance activities in certain areas of the system. Circulars can be added to the billing statement, indicating the timing of routine maintenance or planned disruptions in specific areas.

STREAMLINING EXISTING WORKFLOW. Water and wastewater utilities are often responsible for a wide variety of activities. This may include constructing and managing the development of new facilities and the decommissioning of old facilities. Geographic information systems can aid in managing information related to these capital projects, in terms of physical assets (as-builts) and paperwork such as permits, construction workorder management, and inspections. The idea is to put the information in the hands of the staff empowering them to perform their jobs to the maximum efficiency.

Day-to-day operations can also benefit by leveraging specific features of GIS. Pipeline inspection reports can use dynamic segmentation and image integration features to facilitate the storage and display of sewer inspection images. As a camera moves through a pipe, the distance measurement is recorded at a defect. This information in conjunction with the description of the pipe defect (roots, grease, or other problems) can be captured for further analysis.

The most obvious benefit of these applications is the ability to obtain accurate and timely information about utility infrastructure. Readily available data provide for better and easier decision-making with regard to maintenance and the future planning of new infrastructure.

The efficiency in tracking and monitoring utility operations can, over time, reduce the cost of the activity. Plans are effectively being produced that determine areas slated for rehabilitation, thus reducing the costs of major infrastructure improvements. These programs will also support the prioritization of budget programs for utilities' core activities. With readily accessible and accurate information, drawings can be prepared more efficiently and reduce the time in the permit-review process and approvals. The ultimate benefits are being passed on to the customer through increased productivity and reduced labor costs.

IMPROVED CUSTOMER SERVICE. For wastewater operators, as-builts drawings and reference map sheets are essential for their day-to-day activities. Drafted many years ago, these map sheets are often out of date. Bringing the map sheets up to date can be a major project. For the most part, these are referenced to mapping information such as a building corner or legal property line. The building may have changed and locating the property line can be difficult at times (i.e., in winter). The operator then must guess at the location, potentially wasting valuable time. Geographic information systems can improve the efficiency of the operation by either providing timely mapping information in paper format or displayed on a mobile computer carried by the operator. Combine this with a global positioning system (GPS) receiver and the location can be found easily.

Operations and maintenance personnel make up a critical component of a wastewater utility. The nature of the business can be reactive in that utility employees are often responding to customer complaints and requests. The information required to support the activity must be carried with the staff member and organized carefully. With routine preventative and emergency maintenance and various inspection activities, a workorder management system is warranted. This will aid the operator in locating the problem and arriving with the proper tools and equipment to fix the problem to minimize service disruption. Geographic information systems can track this activity and assist in identifying reoccurring maintenance problems and help in visualizing the planned work.

MAINTENANCE MANAGEMENT SYSTEMS. Maintenance management systems (MMS) can greatly benefit utilities by acting as data repositories and facilitate the development of optimum, multiyear maintenance and rehabilitation priority work programs for each infrastructure component. These can take on various forms and are typically customized to meet specific objectives outlined by a utility. These individual priority programs can then be integrated based on user-defined criteria and an overall priority program is established for all or selected groups of infrastructure components.

Each asset maintained within the management system might use the following analytical process:

- Inventory and condition data collection,
- Determination of current condition and capacity,
- Identification of needs,
- Selection and economic evaluation of feasible alternative treatments, and
- Development of optimum multiyear priority programs.

The integration occurs at the "data" and "user-interface" levels. At the data level, carefully designed database structures facilitate linkages to support management needs.

Inventory and Condition Data. Inventory and condition data are valuable but cannot be used effectively unless they are easily accessed and evaluated and organized for summary and comparative purposes. Building inventory in a computerized database facilitates efficient and effective organization and ease of access by its users. This approach also permits easier updates and expansion. Once data are updated and new data are added, MMS can be used to extract specific sets of data for analysis within MMS or GIS.

When fully implemented, MMS become the primary repositories of infrastructure information. As a result, it is important to ensure that the system is designed and developed for a multiuser environment with user access security, data validation, and data backup–restore features to maintain database integrity.

Data Editing and Display. Allows the user to review and modify asset attribute data. Users who are permitted access security may add new infrastructure components and alter the attributes of items in the database. Through defined user group access levels, other users may only review the data.

Data-Transfer Utilities. The transfer or addition of large volumes of data is facilitated through this system. All data that are added to the system are required to flow through a series of validation and accuracy tests to be sure that no erroneous data are entered to the system. Further, through these utilities, data can be exported to third-party models (such as hydraulic models) for assessment and the results can be imported back to the overall system for further evaluation.

Reporting. Public works professionals, in the access and manipulation of data, can use a series of useful reports. In addition, through an open data structure, clients can use any number of third-party report writers to produce customized, ad hoc queries that more specifically meet client needs.

Multiparameter Condition—Performance Assessment. Maintenance management systems determine the condition of each infrastructure component based on multiple parameters. For example, the condition of sanitary pipes is determined in terms of an overall condition rating that can be based on structural quality (CCTV) and hydraulic capacity of the piping system. Similarly, other conditions of a utility may be determined based on user-definable indexes, such as

- Age or last maintenance cycle;
- Structural condition;
- Hydraulic performance;
- Location;
- Land use;
- Environmental and health considerations;
- Conformance to standards; and
- Disruptive influence, if the utility is removed from the service.

Determination of Needs. Rehabilitation analysis provides the user the ability to assess infrastructure condition, local maintenance and rehabilitation methods, cost, and benefits to determine an overall maintenance, repair, and rehabilitation program based on user-defined budget scenarios.

Through setting a series of need criteria, the user has the ability to generate a complete listing of all infrastructure items that are in need of work. Examples of need criteria could be sanitary pipes that have a collapsed section or storm pipes that are under capacity hydraulically. Based on user-defined need criteria, the system assesses each infrastructure item in the database and determines a listing of those items that are in need. This list is then passed through the following stages to determine the most appropriate program of work to be undertaken.

All potential rehabilitation treatments (from flushing to replacement) are input to the system, with network level costs and benefits, to be used in the determination of activities that could be appropriate to each individual infrastructure item. Further, conditions that would cause each potential treatment to be used are defined.

For each appropriate alternative generated, an economic analysis is conducted to determine the alternative with the highest benefit-to-cost ratio. All alternatives are ranked such that the user may compare alternatives and override selected options.

Based on user-defined budgets (up to 10 years in advance), the module can generate a program for each utility based on achieving the largest benefit to the utility infrastructure for the limited budget.

Life-Cycle Economic Evaluation. For each need identified, feasible maintenance and rehabilitation treatments are first selected through a decision tree process. Life-cycle costs and benefits are then calculated for each treatment to facilitate comparative analyses.

Optimum Multiyear Priority Programs. Maintenance management systems develop optimum multiyear priority work programs for each infrastructure component. Different budget scenarios can easily be analyzed to provide answers to "what if" questions. Annual priorities established can be displayed on a map using an MMS mapping (GIS) component.

Integrated Priority Programs. Individual optimum, multiyear priority programs developed for each infrastructure component are then integrated based on user-input criteria to establish an overall integrated multiyear priority program for the infrastructure as a whole. The integrated priority program, which can also be shown on a map, ensures the most effective use of annual budgets available. Then, MMS assist in the implementation of the integrated priority program by generating appropriate workorders, allocating and scheduling resources, and tracking costs.

CLOSED-CIRCUIT TELEVISION SCHEDULING AND INFORMA-TION BASE. In general, automated mapping and facilities management applications offer an intuitive gateway to information CMMS applications and allow many of the decisions made in the day-to-day operation of a collection system to be greatly enhanced. Scheduling CCTV inspections can be made easier by a graphical representation of what portions of the pipe network have been surveyed. In addition, a GIS system can incorporate information, such as low-flow conditions, that can help prioritize which portions of the collection or distribution networks should be televised. Finally, using the pipe system established on a GIS map as a guide can be used to select video footage of CCTV records. With the advent of digital video applications, this could be a link to an mpeg file on a video server. The user will be able to quickly retrieve the video segment or pictures of the referenced defect in the inspection log database.

PLAN APPROVAL AND PERMITTING. Many agencies are beginning to request that information pertaining to permitting and approvals be received electronically. For this to be effective, data must be in a compatible and concise format. Potentially, agencies will also have a base map for the applicant to build on. With these components in place, efficiencies of the process can be realized. Geographic information systems data typically are accompanied by material referred to as metadata or data about data. This metadata will reflect digital data formats and other compatibility and accuracy limits.

MOBILE COMPUTING. With the advent of the Palm-type computer, mobile computing is no longer a dream; it is a reality. While still in its infancy from the perspective of wireless communication bandwidth, the concept of taking the office with you is not all that far off. The power of data capture is now in the right hands. Quality control/quality assurance can take place in the field. Linking the mobile computer with other applications such as GPS will significantly increase the productivity of data collection. The operator can view his position in real time of near real time. The computer can consolidate information from several sources and present it to the user in an easy-to-understand format. The user has the ability to perform system checks on base data and flag them for inspection or enter a correction. The system can be set up to have some inherent "intelligence", allowing only certain acceptable information in the database. Paper forms are essentially eliminated and data are captured electronically. This significantly speeds the collection and distribution of field data as well as significantly reduces manual data transcription errors. Information can be gathered throughout the day, with a download in the afternoon, or the system may allow for wireless communications if only partial updates are required. This allows smaller data transfer volumes and thus the limitation of wireless bandwidth is overcome.

For GIS to be beneficial, accurate data must be captured and maintained. As utilities turn to GIS and related technologies to integrate data and develop management applications, they are immediately faced with the daunting task of data collection. Although today's computing systems provide the power to organize, manage, and distribute information, data collection—from initial capture to maintaining data integrity—remains the key to these systems' success.

New methods of data capture are constantly hitting the marketplace. With the utilization of GPS location and related sensor data and combinations with any number of attributes defined by the user, building a database for input directly to GIS or the management system is possible. In addition, some applications use speech-to-database software that can help implement GIS and information systems less time than traditional methods, reducing costs, eliminating data redundancy, increasing effectiveness and efficiency, and boosting access to the information that decision-makers need.

Dispatching is also a reality with the combination of GIS and GPS. Locations of problem sites and the closest maintenance truck can be monitored. The inventory of the truck can also be maintained to ensure that it will be able to respond to the call successfully.

PROJECT VISUALIZATION. Through the project-planning stage, GIS expands the possibilities of decision-making by having the ability to overlay and interact with many different data layers of information. Engineers quite often have to communicate with decision-makers that are nontechnical about project elements. More sophisticated GIS three-dimensional models give the user the ability to translate complex data and concepts to a visual experience that communicates ideas easily to the layperson. A final version of any project can be modeled using this system, which is helpful in communicating project elements to the general public.

COORDINATED EMERGENCY RESPONSE. Being prepared in the event of an emergency requires immediate access to the best available information to make quick, accurate decisions. Geographic information systems can help coordinate emergency-related information for various levels of government in a timely manner. For example, federal and state disaster relief offices are developing Web-enabled services to deliver real-time disaster information in a spatial format. In the event of an emergency, an agency using GIS technology can spatially interact with disaster location and severity data with their own facilities to direct resources to where they are needed the most.

The GIS technology has also been used successfully in analyzing hazardous situations such as the storing of hazardous chemicals within the work environment. Inventories of hazardous materials can be spatially rendered and used to assess the risks to workers. In addition, the effects of chemical releases can be modeled with the assistance of GIS to determine the potential effects to plant operators and the surrounding community.

ENHANCED DECISION-MAKING. The goal of GIS technology is to provide users with an abundance of relevant data in an easy-to-use environment. To this end, no technology that has been developed so far can give more to the decision-making process than good GIS. The decision-making process typically involves the evaluation of options—weighing consequences equally until the best solution presents itself to the evaluators. With GIS, the evaluation process can be as simple or complex as it needs to be, correlating relevant information that helps narrow possibilities. This information does not necessarily have to be "linked" in any organized database index. The linking can be facilitated by spatial location and geoprocessing datasets to create a spatial link. In addition, the graphical user interface of GIS speeds the process of decision-making.

CASE SUMMARIES

There are many good examples of how GIS technology has helped the business needs of many organizations across the country. The following case studies offer a few examples of how GIS were used as tools to help organizations meet their needs as well as means for helping manage emergency situations.

COMPLYING WITH FEDERAL REGULATIONS—CITY OF INDIANAPOLIS, INDIANA. The city of Indianapolis, Indiana, has a specialized system of sewers that has come to the attention of U.S. EPA and GASB because of the presence of combined sewer overflow (CSO) rates that are not up to the standards of the Clean Water Act of the United States. Inventory and management of the sewer system in the city has, for years, been either nonexistent or poorly monitored. It is through GIS, GPS, a work management system, and new standards that are now enforcing the management and update of the water and wastewater data within the city that CSO rates are beginning to decline.

Initial efforts focused on field surveys, which led to hand-drafted maps of the system of sewers in and around the city. The creation time for these maps was lengthy and did not have a set numbering system for the management of each of the maps. Because of the absence of an asset-numbering system for the maps, discrepancies were created. Thus, the city had a specific need for asset management for their data and maps.

With the onset of computerized mapping, the city initiated its own computerized mapping program called the SYNERCOM program. This program, like many other early computerized mapping efforts, was without a standardized inventory system to maintain current and updated data. So, again, the city (even though it was converting the paper maps to digital format) was still creating errors that would need to be resolved in the future.

Geographic information systems were then introduced to the market and the city moved from the SYNERCOM program. As the conversion took place, data were misplaced and newer errors were created because of the absence of a complete water and wastewater (sewer) dataset.

With the growth of GIS in the market, new standards were beginning to evolve to improve data quality and management. The city was now under new pressures from GASB to quantify recorded water and wastewater data before the end of 2003. All errors that were in existence at the city had to be resolved within the next two years and data management, through GIS, was about to take charge.

New standards were established to monitor all aspects of the system of sewers in and around the city. These standards, focusing on both the management and quality of the GIS water and wastewater infrastructure, were initially met with opposition and failures to comply by the engineering community and public and private contractors. Meetings and training sessions were held that enabled the city to receive feedback from engineers and contractors to revise the standards to better serve all parties involved, including federal agencies that were monitoring the city's progress. Input was received and standards were updated, but total compliance by all parties was not met at that time.

Data integration and asset management are now allies to the city's efforts to improve data quality and management. Because of the widespread use of GIS in the water and wastewater industry and other industries throughout the world, compliance to data standards is now supported and enforced by law. Thanks to automated procedures using GIS as well as better trained GIS staff, the city finally has a good system for asset management and CSO errors that plagued them in the past (Hammond and Weintraut, 2001).

**NATIONAL POLLUTANT DISCHARGE ELIMINATION SYSTEM—
CITY OF BOSTON, MASSACHUSETTS.** In September 1999, U.S. EPA
issued an NPDES permit to the Boston Water and Sewer Commission (BWSC).
Its purpose was to enforce the regulation of pollutant discharge through catch
basins in the downtown Boston area. The BWSC complied by contracting out
to a consulting firm to develop a customized GIS application and integration
plan that would take existing GIS data and procedures and update them to
correspond to the permit that had been issued. The following case study is a
summary of the implementation of GIS within the BWSC to adhere to the
NPDES permit that was issued in late 1999.

The city of Boston, Massachusetts, has (just like most major cities in the
United States) a system of sewers for water and wastewater drainage. Within
the city's system of sewers are a large amount of catch basins that serve as
filters to prevent obstructive material from entering and blocking the sewer.
Thus, with more than 107 000 people in the downtown area, catch basins have
been known to deplete in quality and release pollutants to waterways. These
waterways are accessible to the public and can cause illness and death if large
amounts of the pollutants are encountered.

With this issue spreading across the United States, U.S. EPA acted to enforce
the Clean Water Act on behalf of improved water quality within municipalities.
A key point of enforcement that relates to the issue in Boston is the NPDES
permit. The permitting process involves evaluating each pollutant discharge
issue on a case-by-case basis as well as evaluating the technology available to
regulate and administer treatment. This same methodology was applied by the
BWSC to implement a program to improve the monitoring system for catch
basin quality in the municipal area.

The initial phase of the project was to establish a strategy for updating the
existing GIS and Oracle database structure. The BWSC contracted out to a
consulting firm to design a new mobile GIS application and integration strategy
that would not only update new and existing GIS data but also build a relationship (topology) between the updated GIS and Oracle database of attributes
(descriptive data for each feature in the GIS).

The consulting firm was challenged with the task of taking an inventory of
all of the catch basins, locating missing catch basins and recording them, and
finally correcting any misplaced catch basin points that had previously been
captured within the GIS. Traditionally, methods of recording data at the BWSC
involved multiple steps of data collection in the field, transfer to paper maps
and databases, and then finally transferring the paper maps to digital format.
The likelihood for error with this methodology was high. Thanks to the catch
basin application (CBA) that was developed, traditional methods were sur-
passed. The CBA enabled the five crews in the field to inventory and map new
and existing catch basins while still in the field. Also, the traditional method
of interpretation from paper maps to digital format was no longer needed due
to having all information available through the CBA's two components (the
GIS and the database).

Once data were collected in the field, the next step was to automate manual
data entry to the Oracle database. To do this, the BWSC used a librarian inter-

face module (LIM) to maintain data within the 277-tile GIS. The LIM enables the BWSC to maintain the data by extracting single coverages to the water network management editor for updates and corrections. This process must occur before data can integrate to the GIS because of internal standards at the BWSC. Thus, the consulting firm, with the help of Geographic Information Technologies, Inc., developed the automated routine to take the updated datasets and integrate them to the Oracle database via "feature" and "facility" identification numbers. Identification numbering issues arose between existing catch basin entries and new entries that were captured in the field. These issues were addressed and reflected in the automated routine through the integration of the catch basins' neighborhood grid location and sequential number when captured.

Therefore, because of the presence of the existing BWSC GIS as well as mobile GIS application and data integration routines developed by the consulting firm, data accuracy and efficiency greatly improved. Furthermore, the output from this project not only helps the city of Boston maintain its pollutant discharge; it also provides an accurate and up-to-date dataset that will be used for future projects within the city of Boston/BWSC (Corriveau et al., 2001).

EMERGENCY RESPONSE—NEW YORK CITY ON SEPTEMBER 11, 2001. On the morning of September 11, 2001, when two commercial airlines crashed into twin towers of the World Trade Center, the City of New York, New York, was prepared for any kind of disaster on a management level. Thanks to the Mayor's Office of Emergency Management (OEM), an emergency response center that could hold and use more than 68 agencies—all focusing on resource monitoring and response during a crisis—New York was in good hands. But as the damage level grew later in the morning and the twin towers fell, so did the OEM because of the danger of fire. Thus, reestablishment of this central emergency response unit was needed. The following case study summarizes the details from an article entitled "New York City—Creating a Disaster Management GIS on the Fly" (Harwood, 2001).

Once the OEM was moved to a new location along the Hudson River (Pier 92), initial efforts focused on gathering data from various sources to provide maps as quickly as possible. When this stage was over in late September, efforts then shifted towards a system that had more regulation of data quality and availability. These data, output in multiple forms ranging from online maps to hardcopy, or paper, maps, allowed key decision-makers, rescue workers, and the public to aid in the response effort and feel safe when they returned to their (temporary and permanent) homes.

A key component to the reestablishment of the OEM was data integration. Because of the nature of the crisis and the number of people affected by its wake, data were scarce and unreliable. Numerous agencies (public and private) and volunteers enabled the OEM to operate again by sending personnel, hardware, software, and data that would shape the OEM into a valuable resource for emergency workers. Even though the OEM's GIS capabilities were initially unknown to workers in the field, word eventually got out and demand for maps increased exponentially.

Sharing the data that were both introduced to and output from the OEM enabled a continuing response effort to grow. Growth took shape through a continuing effort of leadership, monetary support, and field data, as well as a standardized workflow that enabled workers to deliver spatial data (online and hardcopy [paper] maps) in a timely manner. Spatial data were used to give rescue workers updates of building stability, proximity of fire to oil and gas lines, and debris cleanup status at the crash site.

As mentioned before, initial efforts of data acquisition and data output were not as regulated as with normal GIS. But as support continued from all over the country, a need to standardize and synchronize GIS efforts was addressed. Various data sources were integrated (via conversion processes) and then output with a higher degree of data quality. Online efforts were also set in motion to give the public and the mayor's office status checks for the city's emergency response and cleanup progress.

Therefore, all of the response efforts, leadership, data sources, and voluntary personnel were brought together through the flexible and dependable aspects of GIS. But more importantly, GIS helped to make September 11, 2001, a day and an experience that was driven by a sense of humanity to get the job done right and keep people safe. Future efforts to respond to emergencies should, without a doubt, take into consideration the value of GIS and its ability to get up-to-date information to those individuals who can turn disastrous events around for the greater good of the people who are affected by the event.

COMPUTERIZED MAINTENANCE MANAGEMENT SYSTEM— SAVANNAH, GEORGIA. The city of Savannah—Georgia's oldest and one of its fastest growing cities—wanted to create GIS to support information management needs for its water, wastewater, and stormwater infrastructure. A consulting firm worked with Savannah, which had maintained paper maps in the past, to design the Savannah area geographic information system for maintenance and management, system expansion and inventory, water quality analysis, capital improvement planning and budgeting, modeling, engineering and design, reporting, and map generation.

To create GIS with maximum value, Savannah performed a comprehensive GPS utility inventory to update the accuracy and completeness of its infrastructure maps to reflect field conditions. First, hardcopy and digital source documents were georeferenced for use in the field. Field crews then located 47 000 utility features using pen-based, real-time differential GPS, which provided submeter horizontal positions. Real-time mapping was performed, and network connectivity was created in the field using the consulting firm's mobile mapping system proprietary software.

To date, crews found that, of the utility structures located, approximately 1900 had not been recorded on the city's existing paper maps (i.e., features may have been buried, paved over, or hidden by landscaping). Because the inventory revealed the existence of these structures, city maintenance crews can now respond to emergencies and service calls faster and more efficiently.

Crews also collected additional attribute data and recorded condition information by opening more than 32 000 wastewater and stormwater manholes. Thus, Savannah used the inventory as an overall utility system inspection; structures with unacceptable conditions were flagged in the database so that crews could address problems needing immediate attention — problems they otherwise might not have known existed. Additional field inventory benefits include

- Proactive maintenance for clogged manholes,
- Locating and repairing structural damage,
- Features being found and raised (proactive reporting to U.S. EPA and local regulatory agencies),
- Many features being televised (unable to locate from street level),
- Locating utility system interference (sewer pipes in storm manholes),
- Data being used with other projects in the city (database being used with a water complaint table), and
- Data being shared with other agencies (storm information is shared with mosquito control division for West Nile study).

The development of a link between the GIS and Savannah's CMMS will provide additional value over the long run through centralized data management, greater efficiency through workorder processing and tracked work activities, and better maintenance planning.

HYDRAULIC MODELING—GERMANTOWN, TENNESSEE. The city of Germantown, Tennessee, wanted to create GIS to obtain a complete and accurate digital mapping inventory of its water distribution system for facilities management applications and build a GIS demand database to input to the hydraulic model for capital improvement planning.

In the past, the city maintained water data on a $1'' = 500'$ scale mylar sheet. During GIS development, performed by a consulting firm, data on the hard-copy water map were updated, converted, and combined with the city's newly developed digital orthophoto and planimetric base map. This highly accurate GIS dataset could be used immediately for inventory, map maintenance, and modeling applications, and, in the future, for building additional applications.

Once developed, GIS tools were used to extract a skeletonized water network as a data input file for Germantown's water modeling package. The GIS were used to assign customer demand information, taken from the city's billing database and stored in parcel centroids, to modeling nodes. Spot elevations were also extracted from the city's digital elevation model and automatically assigned to modeling nodes. Without GIS, the precise demand assignments would not have been possible.

By using newly developed GIS data to feed the hydraulic model, the city did not have to create a new modeling network. The model was then calibrated and run to determine whether the system could meet present-day demand and where future capital improvements were needed. The city learned that it needed more elevated storage and a larger-diameter line to the feeding tank to maintain system pressures under peak conditions. Engineers are now able to use GIS data to model the system, identify system deficiencies, develop repair-and-replace strategies, and develop a capital improvement plan.

Germantown's GIS also allows information to be managed, maintained, and shared across departments such as public works, police, and fire. The city may soon expand the GIS by adding an online parcel application and a document management system for managing the city's as-built drawings.

REFERENCES

Corriveau, A. R.; Keen, S.; McEachern, J. (2001) Using GIS to Automate Field-Based Utility Inspections and Facilitate Data Integration. http://www.esri.com/library/userconf/procol/professional/papers/pap1049/p1049.htm (accessed 2002).

Hammond, R.; Weintraut, J. (2001) Cleaning Up Indianapolis' Sewer Layers: When Bad Data Happens to Good People. http://www.esri.com/library/userconf/procol/professional/papers/pap425/p425.htm (accessed 2002).

Harwood, S. (2001) New York City—Creating a Disaster Management GIS on the Fly. *ArcNews*. Winter 2001/2002 Edition. http://www.esri.com/news/arcnews/winter0102articles/nyc-creating.html (accessed 2002).

Chapter 3
Geographic Information Systems Planning

*I*NTRODUCTION

Planning is required to successfully implement a geographic information system (GIS). Yet most organizations resist it, maybe because they have planned to excess and have not been able to implement what they plan. Or maybe they believe that GIS is just another way of automating mapping and can be implemented just like computer-aided design (CAD) was a few years ago. Or maybe they see that others are already using their GIS and it will be easy to copy their example.

For GIS to fulfill its promise, it has to deliver benefits throughout the organization, and it cannot do that unless it becomes incorporated to the core business processes of the organization. Planning is required to determine how the GIS will be developed to accomplish this.

Given that GIS implementations often require years to be completed, the planning effort may seem irrelevant to some. But the value of the planning is not limited to the written plan. The real value is in the process, in which members of the organization come to understand how they perform work

---

now, how they can perform their work with a GIS, and what steps are needed to move to the desired future.

In an ideal world, each organization would begin a GIS by following an optimal sequence of events. However, in the real world, it never happens quite that way. A GIS is an investment. Justifying this investment is always a challenge. In many cases, GIS projects are begun to solve specific business problems and have a narrow scope. In other cases, organizations have started to develop a GIS several times, but desired enterprisewide results were not realized. This history means that many organizations may have already acquired some of the elements of a GIS such as hardware, software, data, or staff, independent of a successful planning effort.

MOTIVATIONS FOR GEOGRAPHIC INFORMATION SYSTEMS DEVELOPMENT. There are a number of forces that motivate the implementation of a GIS within an organization. These forces might include

- The need for more efficient operations,
- The need to better serve a growing community,
- The success of a small GIS project that propels the desire to gain the same benefits across the organization, and
- The desire to integrate organizational "silos" through the use of technology.

Each organization will approach GIS with a unique set of existing operational processes and procedures into which the GIS must fit. Because of these differences, planning for a GIS will be a unique effort for each organization. Geographic information systems planning is a flexible process that should be tailored to each situation.

Those who champion a GIS also vary by organization. In some cases, the champion may be someone in an upper management role within the organization. This person has the challenge of convincing operational staff how a GIS would benefit each staff member and the organization. In other cases, the champion may be a staff member. This person has the challenge of convincing upper management that the investment will provide benefits.

This chapter presents a process that has been used to successfully plan and develop GIS in multiple organizations. The discussion will begin by defining the lifecycle of GIS and how planning fits into that cycle.

GEOGRAPHIC INFORMATION SYSTEMS EVOLUTION THROUGH PLANNING. Organizations want a GIS for its ability to help solve problems, save money, reduce redundancies, and increase efficiencies. The ultimate use for a utility GIS is to better serve the customers. Yet when a GIS fails, the most expensive components in creating the GIS—the technology and the data—are blamed: "The computer program had bugs," or "The data wasn't as accurate as we needed it to be". In reality, failure is generally caused by human factors. These human factors include a misunderstanding of GIS capabilities and a lack of confidence in the data's integrity and usefulness.

Indeed, people must plan first and purchase last. In fact, planning will define and drive the entire GIS process, including the eventual hardware, software, and data decisions. The following needs are universal for any organization to prepare for strategic GIS planning and support implementation tactics:

- Provide leadership and support. Find champions throughout your organization to provide leadership and support for new technologies, development, direction, and technical training, and support activities.
- Geographic information systems education. The leadership should meet with other department heads and decision makers to discuss what GIS is, how GIS will fit into the organization, the goals of the GIS and its mission statement, and how GIS can support their function.
- Create requirements for the GIS. Define what the organization expects of this GIS and communicate that expectation. Key considerations include the identification of the scope of implementation (small application versus full-blown enterprise) and integration issues with other systems.
- Create documentation. Develop information to ensure proper use of GIS data layers.
- Implement the plan. Organizations can uncover all of the issues that surround GIS and document its GIS needs before attempting to define the solution. A formal process is necessary and it can be accomplished in many ways.
- Investigate procurement options. Can the GIS be built using only internal resources? Does the organization have the skills, knowledge, and budgetary resources to do this? Or, would it be a wise decision to outsource some or all of the GIS implementation and development activities?
- Budget and plan for the GIS. A GIS will require a significant financial investment, and the return on the investment is not always obvious during the planning process. What will it cost in dollars to build and implement the GIS and what will it take to justify this investment?

A GIS constantly evolves. There is no ultimate "maturity" level. As an organization grows, so will its GIS. Addressing the universal needs above will place a foundation with the organization to plan, implement, and grow GIS.

Even though each organization's GIS implementation is different, there is a logical sequence for GIS implementation. The following sequence generally follows an information technology (IT) endeavor. However, there is one significant difference—the need to develop data.

- Planning—the stage discussed in this chapter.
- Design—during this stage, detailed design documents are developed for data conversion and application development. Also in this stage, the database design is typically completed.
- Data creation and conversion—typically the most expensive stage.
- Core technology acquisition—purchase and installation of networks, hardware, core GIS software, and so on, needed to implement the GIS.

- Application software purchase and development—purchase or development of application software to solve the organization's specific business problems.
- Deployment—deployment can be in many stages, and each organization must prioritize deployment depending on the most important needs of the organization.
- Maintenance—the ongoing commitment to maintain the data, systems, and applications over time.

PROVIDE LEADERSHIP AND SUPPORT

Organizations implementing GIS are diverse in nature and exhibit individual behaviors and structures. Even those organizations belonging to a specific niche, such as water and wastewater, often display a broad spectrum of characteristics. They range from small companies to large corporations. They are associated with both the public and private sectors and just about every variation in between. As would be expected, individual management and organizational structures, coupled with varied leadership and support cultures have a significant effect on planning and development for GIS.

INFLUENCES OF SUCCESS. The success or failure of GIS is strongly influenced by the strength of leadership within an organization and by the overall support structure within the organization. Leadership and organization present a set of boundaries in which GIS planners must operate. These boundaries may be closely aligned with the departmental and hierarchical structures of the organization. Or these boundaries may be more closely related to individual support, for example, a managerial champion. Geographic information systems planners should have open dialog with other department heads to educate them to the needs and value of GIS. This is an opportunity to promote GIS and the benefit it can bring to the organization.

Where a GIS manager or planner falls within the structure may also play a key role in how effective planning is formulated, presented, approved, and implemented. The individual tasked with planning the GIS is often viewed as the leader and this person is commonly a catalyst for identifying the support needed for GIS development or is the source that fosters the growth of the GIS.

AUTHORITY AND RESPONSIBILITY. The position occupied by the GIS planner in the organizational hierarchy may significantly affect how plans are formulated and implemented. The organizational structure typically provides the basic framework for accountability, authority, and responsibility.

Particularly in situations where GIS has been introduced at the departmental level, a dichotomy may exist between the responsibility of making planning decisions and the authority to adopt and implement them.

Typical GIS managers may have the responsibility to establish standards for GIS-related hardware, software, and data standards but they may lack the authority to formally adopt or implement those standards companywide. Additionally, a GIS manager may be responsible for purchasing GIS-related hardware and software, but the manager may be limited in budgetary authority over established spending amounts. This allocation of authority is not a bad practice; in fact, it is an essential part of any organization's checks and balances. However, it does mean that effective planning must recognize organizational limitations and the way they apply to the GIS manager's role.

The migration of many organizations from historically hierarchical structures to "flatter" structures where individuals are empowered to make decisions has the potential to exacerbate the authority and responsibility conflict as control boundaries become blurred. Additionally, GIS crosses these organizational boundaries and it is often difficult for authority to follow suit.

Insufficient authority blocks progress. The GIS planner without adequate authority is forced to rely on persistence to accomplish objectives. A strong understanding of the organizational authority and responsibility boundaries will allow a GIS planner or manager to present goals and objectives in a manner and in a format that is expected by the organization's decision-makers and identify areas where additional support strategies may be necessary.

EXPECTATIONS. Planning may produce several outputs, including development and implementation plans, system upgrade recommendations, and project cost estimates. Effective planning garners support by tailoring these outputs to the expectations of decision-makers. Planners in differing circumstances will have different expectations. A board of directors may require different information than an agency director, city and county commissioners, or chief executive officer. A planner may have only one or a variety of such authority layers with which to interact during the planning stages.

While identifying and understanding expectations is somewhat reactive, the effective planner must sometimes proactively play a role in setting the expectations. There is a delicate balance involved in developing expectations that are realistic, yet optimistic with potential. This means educating decision-makers and providing them with a generalized high-level overview of what to anticipate as part of the coming GIS planning agenda. Using this approach decreases the likelihood that decision-makers will be surprised by GIS initiatives and encourages recognition that GIS planning is an incremental, ongoing, and necessary process.

SUPPORT STRATEGIES. One method of setting and defining expectations is to formalize the planning pathway. This means developing an organizational awareness that GIS technology has the potential for broad positive effect and, therefore, GIS planning should be an expected and integral part of the overall organizational planning process. Typical implementations of such pathways involve the formation of technical and policy committees to address appropriate issues. As their names suggest, these committees provide a team-based mechanism for communicating and addressing issues related to different aspects of GIS planning and development.

Technical committees typically foster involvement and ownership from individuals who work with the data and associated systems. These individuals are familiar not only with GIS idiosyncrasies, but with its effect on everyday organizational work processes.

Policy committees frequently address high-level policies and procedures and may even act as the structure by which GIS planning is formally linked to other organization and information technology goals and objectives. The policy committee may act as the final layer before planning initiatives are submitted to decision-makers, or the committee itself may be the group to make final decisions. The degree to which such formal committees are necessary is often related to the size and complexity of the organization itself.

In organizations that are reluctant to adopt these support structures, the GIS planner should identify and communicate the risks of disjoint planning or development that is not integrated with other organizational and information technology efforts. Some of the most significant risks are

- Duplicated effort: inefficient use of personnel and monetary resources when goals and initiatives from separate projects overlap;
- Untimely development: applications and systems that do not function properly or to full potential because key building blocks did not exist before development (this is especially hazardous for systems integration); and
- Incompatible systems: applications that are incapable of sharing data or cannot be easily integrated because of different design architectures.

Mitigation of these risks is a significant incentive to coordinate planning efforts. Regardless of the specific method of implementation, formalized support structures have considerable potential to foster support for GIS at multiple organizational levels and help clear the path for GIS planning priorities. In addition, support from decision-makers is a significant step toward ensuring that GIS planning and development is closely aligned with organizational strategic objectives.

ORGANIZATIONAL STRATEGIES. As a growing number of organizations embrace the concepts of strategic planning, additional opportunities become available for integrating GIS to a more holistic planning mentality. The key is often identifying and solidifying relationships between achieving corporate or organizational strategies and GIS development. This is yet another opportunity to put GIS development needs in a context that will be understood and expected by managers and other decision-makers. Obermeyer and Pinto (1994) identify a clear relationship between organizational strategic planning processes and the mechanisms that lead to successful GIS technology planning. Strategic planning is future-directed and flexible, with an emphasis on being able to identify, compensate for, and evolve with change. This mode of thought meshes well with the dynamic challenges posed by changing IT requirements. Therefore, a GIS plan that is formulated with a strategic planning methodology is likely to have many elements in common with overall organizational strategies. Linking these common elements is a critical step in aligning GIS development with the organization's core mission and goals.

The GIS planner should also be aware of the level of detail in the organization's strategic plan. There are wide variations on how specific goals and strategies may be depicted, and the details concerning how an organization intends to reach specific objectives may get lost in generality. At the organizational level, GIS is not typically a goal unto itself but a tool that will enable a company or agency to realize goals. A water utility may have a goal or strategy for improving monitoring of water quality. If GIS technology is expected to play a role in this strategy, that role should be identified with other substrategies or tactics. When specific tools are not identified and linked, there is an increased risk of losing the connection between GIS plans and organizational plans. These substrategies and tactics may serve as a direct reference to strategic plans geared specifically for GIS and IT, where technical specifications and resource needs can be outlined in appropriate detail. In this manner, strategic planning can help coordinate both the "macro" and "micro" perspectives of GIS development.

ORGANIZATIONAL STRUCTURE. Effective leadership and support may also be affected by where GIS management and resources are positioned within the organization. Every organization has its own GIS history, and that history is often shaped by the method in which GIS was first introduced to the organization. Geographic information systems may have been primarily developed as the result of a high-level mandate or it may have been initiated as a part of some specific project or application. Both successful and failed GIS implementations exhibit these roots and many others.

Utilities, in particular, have often experienced conflicts between the engineering-related aspects of GIS and those associated with IT. In this regard, there are typically three philosophies:

(1) A GIS should be managed from within engineering to build on the foundation of the system being tightly associated with the source of its data. Engineering is also where initial GIS expertise often resides.
(2) A GIS should be managed by the IT department to be more closely aligned with system-integration standards and be treated as a "core" technology.
(3) A GIS should be managed by a dedicated GIS department to facilitate a true enterprise development focus and create an environment for appropriate authority and responsibility. For organizations establishing a GIS with significant size and complexity, this organizational structure may even include a geographic information officer to adequately manage GIS planning development for the whole organization.

The inability to place well-defined organizational boundary conditions around GIS technology is a function of its inherently multifaceted nature. Because no one solution fits all, the best recourse is often not to try and solve the dilemma but to recognize the need to work around the barriers as the system grows.

Regardless of the situation or department, effective GIS planning and development can be achieved if the leadership and support structure recognizes several broad premises:

- A geographic information system is, by its own name, an information technology that should, where possible, adhere to general IT standards to promote integration of data, system architectures, and associated support.
- A geographic information system is not *just* an information technology. It is a spatial technology, subject to its own additional set of rules, standards, and expertise that will determine how accurately and effectively the system will be designed and function.
- Development of the GIS is best managed by individuals or teams having an intimate knowledge of both the IT and spatial aspects of GIS.

It is realistic to expect that, regardless of the manner in which GIS is introduced to an organization, change will occur over time. As the tool grows beyond departmental implementation into more "global" applications, the perspective from which planning and development decisions are being made must also evolve. The customers and products will change. The need for well-defined leadership and support structures will increase, and individuals who manage and plan GIS must be in appropriate positions of authority to develop appropriate controls and support organizational needs.

MULTIAGENCY EFFORTS. Geographic information system planning for an organization frequently involves sharing data and coordinating development with other agencies' and utilities' demands and is both an inward and outward focus that introduces additional dynamic layers to the planning process. Participants in a GIS consortium are typically entities with expressly different purposes. They are city and county governments. They are water, wastewater, and electric utilities. They are parcel mapping and tax-assessment agencies. These diverse organizations have diverse agendas, fiscal year cycles and budget constraints, and levels of expertise and commitment.

It is a somewhat daunting task to bring these entities to the same table and reach consensus on matters that may affect the return on GIS investment for all involved. Often, one agency must take the lead in pushing or pulling the other participants through this process. Consortiums may find it necessary to use some of the same support structures used to facilitate planning for individual organizations, such as technical and policy committees. Successful multiagency efforts depend on developing a mutual understanding, where the consortium is viewed as a necessary and expected part of the individual organization's planning process and where the participants are recognized as valuable components of the consortium.

KNOWLEDGE AND OUTSIDE CONSULTANTS. The GIS planner must be able to lead the administrative aspects of GIS and the technical aspects. Unlike an engineering design project, a planning effort is largely interactive— the need for knowledge exchange is high. A geographic information system is a continual learning process and it is natural to locate as many supporting sources of knowledge as possible. In this capacity, planners call on their own knowledge base, the knowledge of staff and others around them, and often knowledge from external sources such as consultants.

In most utility organizations, initial GIS planning will be a one-time activity. Because of the magnitude of this effort and the knowledge base required, the GIS planning process typically involves the use of a consultant. There are many reasons to use a consultant:

- Consultants have a breadth of experience that they can provide a utility because of their exposure to many organizations and each organization's unique experiences,
- Consultants will have strong GIS technical skills and expertise,
- Consultants can be an independent voice in dealing with organizational and managerial issues,
- Consultants can add skills that the organization does not possess, and
- Use of the consultant's staff can mitigate the staff resource effects on the organization.

There is, however, no substitute for building and maintaining an internal knowledge base. The GIS planner and staff must be in a position to continue where the consultant leaves off and continue planning incrementally while using consulting assistance on an as-needed basis.

CREATE THE REQUIREMENTS FOR THE GEOGRAPHIC INFORMATION SYSTEM

Implementing a GIS is complex. Geographic information system projects evolve through the following phases: investigation, planning, conversion, application development, implementation, and production. No one doubts the need for planning a GIS implementation. The basic principle of project management is coordination that can occur only if there are commonly held goals and organized activities to achieve those goals. To undertake the development of a GIS, there is a need to accommodate multiple goals and participants, to establish priorities, and gain coordinated effort through the phases of data conversion and developing applications to meet user needs. In larger projects, significant subprojects such as data conversion become self-standing, with a defined goal, schedule, resources, and scheduled endpoint. All of these activities are driven by the common goals and action plans articulated in the implementation plan.

ORGANIZATIONAL REVIEW. Planning begins by taking an objective look at the organization. To assist with this review, select key individuals. Ideally, these individuals should have decision-making authority or are those who have the most to gain from technology integration through the organization's operations, maintenance, mapping, engineering, and finance departments. This is an opportunity to not only construct a GIS, but to also build bridges among departments. Forming a team can be the beginning of a number of positive organizational changes. The team, while performing an organizational review, will examine core components of the organization's information environment. These components include

- Culture—values and beliefs about information,
- Behavior and work processes—how people actually use information,
- Politics—pitfalls that can interfere with information sharing, and
- Technology—what information systems already exist.

Culture. A culture can be fast- or slow-paced, relaxed or conservative, technology-oriented or clinging to old-fashioned methods. A fast-paced, relaxed, technology-oriented culture mostly likely will embrace the exchange of information initiated by a GIS. The opposite kind of culture—historically more protective of information—may counter the kind of information-sharing activities that are needed for a successful GIS.

Culture is also about personalities and the way people make decisions. Some individuals are good at making decisions because of an inherent sense of vision. Others are not so confident. They may need guidance and assurance.

Behavior and Work Processes. It is important to understand and categorize the desires and needs of the organization. Items that staff may desire are not always the items that are needed to solve a particular problem. By effectively identifying and categorizing items that are needed today and are desired for the future, GIS can be effectively implemented at the organization.

Ideally, information flows in a linear fashion from "point A" to "point B" to "point C". And at each point along the way, the information is used in specific ways. In identifying how information technology and the GIS will strategically affect behavior and work processes, consider how information flows from department to department, or not at all, and whether bottlenecks are caused by people, processes, lack of technology, or all of the above. Determine whether employees will use GIS technology to do their jobs. Also consider what kind of behavior modifications and training programs will be needed to transform staff.

Assuming that you have a GIS, "point A" could be a technician in the field who is using a laptop computer to record the exact location of a main break in the GIS. "Point B" could be a customer service representative who is using the GIS to communicate to customers affected by the main break. "Point C" could be a department supervisor who uses the GIS to prepare a report about how the main break affected customers in the service area.

In a perfect world, A, B, and C are computer-savvy individuals who are able to access and share the same data. In the real world, the field technician may be using the GIS and even the laptop computer for the first time. The customer service representative may bypass the time-saving and more accurate GIS in preference of paper maps, simply because it is more comfortable using "the old way". And the department supervisor may not have computer access to any of the information input by the maintenance or customer service departments.

Politics. An organization is affected by external politics—the priorities that are set by elected officials and their constituents. And an organization is affected by internal politics—the priorities of various departments.

Externally, elected officials with progressive platforms are generally more supportive of technology efforts such as a GIS. Organizations with officials who see no need for a GIS or whose officials do not understand the uses and benefits of a GIS may be denied the monetary and information-sharing support that would be needed to launch a GIS.

Meanwhile, there may be competition among departments internally for money from an already slim budget. This reduces the chance of funding a GIS. Things may even get personal. For example, one department may distrust another department because of an incident that occurred in the past. Distrust and personal issues can impede information sharing.

Technology. Naturally, it is necessary to review existing hardware, software, and data. This element of an organizational review indicates gaps in the information infrastructure and allows for the development of an appropriate budget.

When examining existing technology, one should keep in mind that the key is not only the technology itself, but also how to accommodate changes in technology. Hardware and software quickly grow obsolete. The GIS plan must accommodate a GIS industry that is dynamic.

Once the organizational review is complete, answer the following questions:

- Will the organization's culture, behavior, work processes, politics, and technology allow for a GIS?
- Is it desirable for a GIS to change the organization's culture, behavior, work processes, politics, and technology?

Indeed, some organizations use a GIS as a rationale to change their entire structure. In this case, the GIS is not merely a technological end product but is then the driving force behind a new way of conducting business.

Planning Overview. A typical planning process has three fundamental steps

- Step 1: perform a needs analysis,
- Step 2: develop a vision and strategy, and
- Step 3: develop an implementation plan.

The components of each step can vary based on the specific needs of each situation. A typical approach is described below.

PERFORM A NEEDS ASSESSMENT. Interview Process. Those who do take the time to plan for a GIS typically start here. A needs assessment is a look at the kind of GIS that an organization requires. By its definition, a needs assessment is a document that tells *what* is needed rather than the specifics of *how* to get there. The needs assessment is a crucial step in the planning process, but it cannot be overemphasized that a thorough organizational review must precede the needs assessment. The *how* is a part of the implementation plan for the GIS as can be seen in Figure 3.1, which shows the spiral model of GIS planning.

Figure 3.1   The spiral model of GIS planning.

The needs assessment helps identify key goals at the outset, keeping the GIS on track. The track is a spiral that starts small, then swirls and spreads out. This spiral strengthens as new issues are identified and incorporated to the GIS plan. The spiral model of GIS development allows an organization to remain flexible and open to new ideas. With this model, an organization will get started on the GIS (or any information system, for that matter) before every question is answered in detail and decisions are permanent. The spiral model discourages an overplanned and static GIS in favor of one that is fluid and dynamic.

The first task is to conduct workshops to interview the department heads and staff most likely to use a GIS about the issues they face. Engineering issues can include water loss, water quality, sanitary sewer overflows, and flooding. Operational issues can include aging infrastructure, ongoing maintenance, capacity overloads, customer service, and staffing. Management issues can include funding, privatization, and government mandates, including Capacity, Management, Operation and Maintenance; the National Pollutant Discharge Elimination System; and Governmental Accounting Standards Board statement number 34. An organizationwide issue may be "information redundancy"— each department acting independently, creating and maintaining duplicate data while not realizing it.

Secondly, consider touring organizations that already have a GIS in place. Take care, however, not to copy someone else's GIS. The danger here is seeing someone else's successful GIS and emulating it. No two organizations are alike and no two GIS developments are alike.

As you tour organizations, ask these questions

- How did you begin the GIS process?
- How does it continue today?
- What have been the pitfalls?
- What have been the successes?
- What are some of the issues you face?
- How is the GIS solving these problems?

Technology Review. Upon completion of the interview process and tours, it is typically discovered that a GIS is really just the beginning of a comprehensive information infrastructure.

A geographic information system is a blanket term for an overall information technology solution in which multiple systems may be supported by individual databases or from one central repository—a data warehouse. Many different users may access the information from these data environments. Multiple systems feeding these data environments are

- The GIS itself—the GIS is a collection of spatial, digitized maps. These maps can be created through a variety of methods: original source documents, global positioning system (GPS) satellite surveying technology, and orthophotography.

  Geographic information system functions include better map maintenance, relational database management systems administration, and engineering document management. The GIS component allows spatial analysis of graphic and nongraphic data to be contained within and connected to the maps. This then allows the GIS to connect to other functions such as

- A computerized maintenance management system (CMMS)—the CMMS allows for improved in-the-field maintenance and construction efforts.
- The customer information system, which aids in customer service, billing, and dispatch.
- The facility accounting system, which aids in inventory management and project tracking.

Of the above technologies, assess which are needed to resolve organizational issues and whether they use common information.

- The laboratory information management system, which aids in water sampling and reporting.
- The supervisory control and data acquisition system, which allows for real-time monitoring, water-loss and trend analysis, and nonpeak pump savings.

But a GIS is more than a technology wheel with a database at the center. It is also a series of stages. The first stage, of course, is planning, the topic of this chapter. The second stage is computerized inventory, in which data is collected and the GIS is built. An organization should assess whether the GIS needs to encompass all six stages to resolve organizational issues. The third stage is spatial data dissemination, in which intranets and the Internet are often fast, easy, cost-effective, and secure ways to share data with in-house staff and with the public. The fourth stage is information management, in which the GIS is linked to one of more of the "wheel" technologies. The fifth stage is modeling and data analysis, in which GIS data is used in conjunction with modeling programs to depict and predict scenarios. The sixth stage is strategic planning, in which GIS and modeling data are used in capital improvements planning. In most environments, the complete evolution of the GIS is the ideal situation. However, every organization is different, and each must consider its own needs in deciding how far to take the GIS.

In terms of data, a few things affect the types and accuracy of the data that are needed. These include technology needs, the stage at which the GIS will be brought in, and design constraints such as a high groundwater table or a hilly terrain. It is also important to keep in mind that it is useful to collect only the data that an organization is willing and able to maintain. An organization must decide the intended current and future uses of the data and the driving forces behind the accuracy of the data.

Naturally, the amount of funding available for a GIS forces a prioritization of technologies and data. A cost analysis can help in this process. And it can be difficult to estimate direct and quantitative benefits associated with information technologies. Because of this, a cost–benefit analysis may be impractical to perform before launching a GIS. However, it is recommended to perform

---

this analysis before purchasing hardware, software, and data. The following is a simplified version of what, in reality, is an in-depth process:

(1) Obtain cost estimates from hardware, software, and data suppliers, as well as from GIS consultants.

Determine how funding limitations impede the development of the GIS and how might the GIS implementation and related information management systems integration be restructured to achieve the most value for the dollar in short- and long-term scenarios.

(2) Evaluate each technology on the GIS "wheel" for its costs and perceived benefits to the organization.

(3) Evaluate each stage, from computerized inventory to strategic planning, for its costs and perceived benefits.

(4) Evaluate the costs and future data uses of highly accurate versus moderately accurate data.

DEVELOP A VISION AND STRATEGY. At this step, the organization will discuss the needs that were identified and begin to determine where technology can be applied to improve business processes or services. Needs are prioritized through a voting session.

Goals for this task include reviewing the existing conditions and discussing and presenting potential recommendations for these topics

- Organizational topics (staffing, training, and so on);
- Hardware, software, and networking resources;
- Data standards, data development, and data sharing;
- Application development and data maintenance;
- Opportunities for system integration; and
- Schedule and budget.

DEVELOP THE IMPLEMENTATION PLAN. Developing an appropriate implementation strategy requires a thorough evaluation of all information acquired to date. Certain plan components, schedule, technology support, data availability, and the coordination of user needs may indicate substantial benefits to developing alternative approaches.

Implementing a GIS is a continuous investment in three areas

- Technology,
- Human resources, and
- Data development.

In addition, a GIS should be viewed as a program that needs to be addressed the same way as other annual budgetary line items. A successful implementation strategy develops the foundation to meet the immediate goals to create current GIS data layers and distribute these data both within and outside the organization. The following items are requirements for a successful implementation:

- **Recommendations** that are phased over time so that they can be appropriately budgeted.
- **Resource** issues that are addressed to ensure that the appropriate mix of internal and consulting resources are applied to the development.
- **Data development** issues that are identified, with a focus on end results and benefits. Also, all necessary baseline information must be obtained.
- **Budgets and funding** that are adequate and allow for any changes in technology that will occur over the lifetime of the project.
- **The management structure** that is identified and developed to facilitate the development and implementation of GIS within the organization.

Strategically, implementation is typically divided into three phases

- Design,
- Pilot, and
- Full implementation.

Design Phase. The design phase creates the framework of the core GIS. This typically includes developing an enterprise database design to include all relevant utility layers. A plan to convert the layers is also defined. This involves specifying the requirements of a utility network development—for example, whether the network is developed using GPS methods or as-built data conversion methods and specifying the levels of accuracy that are required. Initial staffing and organizational changes need to be put into place during this phase as well. The success of the project depends heavily on the availability, dedication, and training of the human resources that are required to perform this project. Generic data models exist and may be applicable for some organizations. The suitability of the data model will vary based on the complexity of the organization's needs.

Pilot Phase. A pilot phase typically focuses on testing the database design and conversion procedures defined in the previous phase on a small portion of the organization service area. A pilot phase defines, tests, and then refines data collection and conversion efforts. A pilot area is selected—targeted sections of the organizations service area will be chosen that best represent anticipated data collection and conversion problems—to use as a starting point.

This phase results in utility and data layers. Data creation will involve the following efforts:

- A field inventory of utility structures and
- Digitizing utility maps based on collected field data or converted source documents.

During this phase, applications are deployed to view, query, and map converted data. These are often Internet or intranet based. The applications

---

can be expanded to support additional functionality or can be made available to a wider audience of users.

Full Implementation Phase. The full implementation phase continues the work of the pilot phase. When the pilot phase is completed, the organization typically evaluates the data that were collected and converted and the methods and applications that were used during the pilot phase. The conversion process and application development efforts can then be adjusted to best meet organizational needs.

Full data conversion activities occur during implementation and can extend over a number of months or years. Additionally, during this phase, the GIS is integrated with other systems, including external systems such as a CMMS or modeling systems.

It is recommended that organizations perform an annual review to reprioritize development activities because they can involve significant expense and effort. Recommendations presented for this phase can be implemented to accommodate budgets and available resources over a period of several months or years.

*I*MPLEMENT THE PLAN

As described at the beginning of this chapter, planning is a process. Planning must be approached as such to deliver successful outcomes. This section presents a detailed process for developing an implementation plan. It can be adapted to meet the specific needs of almost any utility organization.

For this section, it is assumed that a consultant will be used for the planning process. This consultant may be one or more internal staff or a more traditional outside consulting firm. In either case, the consultant must have a good understanding of the planning process as well as a good understanding of GIS.

THE PLANNING PROCESS. Planning follows a specific process that is designed to lead to a successful plan. The process is typically presented as follows:

- Perform an analysis of the needs to be met and the problems to be solved,
- Define the a shared vision of the goals of the solution, and
- Define the specific action steps needed to accomplish the goals.

This is a generic process and can be successfully accomplished in many ways. However, in following a formal process, an organization can uncover all the issues surrounding the GIS and document all of the needs before attempting to define the solution. The detailed process presented here is an adaptation of the generic planning process to the specific needs of a typical public utility GIS setting.

Step 1: Perform Needs Analysis. *PREPARATION*. As described above, the planning process can be performed by in-house staff with expertise or by outside consultants. In either case, some preparation is required before the formal initiation of the planning process.

Prepare to begin the planning effort by

- Defining the exact scope of the planning effort,
- Obtaining commitment from affected staff to participate, and
- Setting up initial meetings and interviews.

INITIATE THE PROCESS. Begin the process with a team kickoff meeting to bring all of the players together and define common project goals. If possible, each key decision-maker for each department affected by the GIS project should be encouraged to attend.

The kickoff meeting will typically include the following elements:

- An introduction to GIS (if needed)—"GIS 101";
- An introduction of project objectives;
- An introduction of the project team and an explanation of team members' roles and interests in the project;
- An explanation of project approach and schedule;
- An outline of the interview process;
- Excitement for GIS with examples—discuss relevant real-life projects to bring a vision to the organization of what GIS can do; and
- Identify as many individual goals as possible—find out what each department expects.

This meeting should do more than explain the project—it should generate excitement for the benefits of GIS to the potential users so that they will be motivated to participate actively in the project. This can best be done by showing examples of successful GIS in operation in other, similar organizations.

Generating excitement for the GIS by communicating the benefits at the kickoff meeting is also a good forum to identify departmental goals in a group setting. The GIS planning process involves many individuals from many departments, so it is natural that there can be a great deal of misunderstanding or disagreement about the goals of the GIS. The kickoff meeting can identify those departmental goals that fit in with the GIS goals. A technique that can accomplish this is to have each participant, in turn, to identify one key goal he or she may have for the project. This does lead to obvious goals that all can agree on, but it also leads to identification of some goals that do not fit well with the project. This process starts a consensus-building exercise that is important to the success of the GIS.

INTERVIEW PARTICIPANTS. The consultant will need to interview key staff members. The number of interviews depends on the size of the organization; however, all upper- and middle-management staff should be included in the interviews as well as all staff that are directly involved in information systems and mapping operations.

Interviews are focused on the use of spatial information and the activities that relate to its development. Additionally, these interviews serve to document what information flows within and among departments and identifies the paths of how the information flows among departments and individuals. Developing a thorough understanding of the "official" and "unofficial" ways that information is shared throughout the organization and the way work is performed is crucial to a successful GIS implementation.

A challenge during the interviews is to capture an appropriate level of detail. All key work processes and supporting information systems need to be identified during the interview process. But detailed workflows of each work process are not needed during the planning stages. As long as significant work processes are identified, additional detail can be captured during subsequent stages.

As part of this task, the consultant should create a series of diagrams to document existing work processes and information flows. A typical workflow diagram is shown in Figure 3.2.

EVALUATE EXISTING PAPER, COMPUTER-AIDED DESIGN, AND GEO-GRAPHIC INFORMATION SYSTEMS DATA. While the interview process is occurring, existing spatial data from all sources, both internal and external, will need to be evaluated. Spatial data could include

- Design record drawings;
- Utility atlas maps;
- Field inspection reports;
- Manhole, valve, and hydrant inventory records;
- Infrastructure inventory databases;

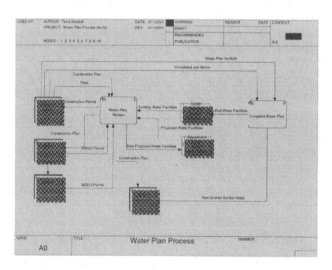

Figure 3.2 Typical workflow diagram.

- Infrastructure asset information stored in maintenance management systems;
- Infrastructure asset information from hydraulic models;
- Existing GIS software and data;
- Imagery (aerial photography, orthophotography, etc.);
- Compiled planimetric and topographic mapping; and
- Cadastral (parcel) mapping.

The following information should be compiled for each distinct class of data:

- Geographic extent,
- Creation and conversion date,
- Method of creation,
- Scale,
- Accuracy,
- Completeness,
- Currency,
- Number of records and sheets, and
- Contact information.

EVALUATE EXISTING INFORMATION SYSTEMS. It is also important to assess the relevant information systems that are currently in place, including applications and databases that were developed internally, off-the-shelf applications, and existing relational databases. Key information should be captured for each data table, including

- Table name,
- Where information is stored,
- What information is included,
- Number of records,
- Maintenance status,
- Owner or maintainer,
- Users,
- Key fields, and
- Linkage fields.

Assess core computing hardware technology, including desktop personal computers, servers, plotters, printers, scanners, and other peripherals. Each component should be inventoried for performance indicators (processor speed, memory, disk capacity, throughput), the date the computer was installed, connectivity, and condition.

Analyze existing local area networks and wide area networks to determine capacity to handle the large data sets that a GIS will impose. This analysis should include development of a network diagram and identification of the available bandwidth of each network link. In some cases, it might be necessary to conduct network analysis if there is some question about the available bandwidth. Network protocols must also be documented to gain an understanding of the ability of the network to support GIS connectivity.

COMPILE NEEDS. Summarize the information from interviews and investigations into a needs assessment summary table. Information that is contained in this needs assessment summary table will be used to develop a prioritization for the GIS in the visioning stage.

PREPARE DRAFT NEEDS ANALYSIS REPORT. The information gathered so far should be used to prepare a draft needs analysis report that documents the findings. This report should be designed so that it will become the initial sections of the overall project report. It is important to stop and prepare the report at this time so that the information that has been gathered can be compiled together and reviewed before it is used for decision-making purposes. This report will typically include two primary sections

- Existing conditions—this section documents what has been found but expresses no opinions other than those presented by the organization's staff.
- Needs analysis—this section states the needs in detail, grouped, and categorized.

MEET TO REVIEW THE NEEDS ANALYSIS REPORT. All key participants in the project should meet to review the draft needs analysis report. If it is discovered that key issues have not been adequately explored or are not understood by the team, more interviews may be needed to fill in the gaps.

UPDATE NEEDS ANALYSIS REPORT. The comments from the review meeting must be incorporated to an updated version of the needs analysis report. Unless this document varies greatly from the initial version, another review cycle can generally be deferred at this time.

Step 2: Define Project Vision. As discussed in the previous section, the success of the GIS requires the development and adoption of a shared vision of what the GIS will be and the way it will be used within the organization.

PREPARE GEOGRAPHIC INFORMATION SYSTEMS FUNCTIONAL REQUIREMENTS ANALYSIS. Initially, the consultant will evaluate potential GIS applications for the organization's use. The initial applications are those applications that have the greatest potential effect for the organization for the least amount of time and money. This approach is used because it is impossible to implement all of the applications that will be identified—at least immediately. The initial applications become the focus of the organization's plans and provide goals that can be accomplished. Strong support and buy-in from executive management and the user community typically follow.

Some of the potential GIS applications that might be included are listed in Chapters 1 and 2. Additional applications may arise from the needs assessment process.

For each application, the consultant should list the application's potential users, data needs, benefits, and cost components. The consultant will compile this information to create an organizationwide GIS functional requirements section of the project report.

EVALUATE TECHNOLOGY ALTERNATIVES. The consultant should investigate and evaluate a number of potential technology alternatives to provide the GIS data management solutions. Among issues to be investigated are

- What organizational structure is best suited for successful implementation of GIS?
- Should the organization standardize entirely on one vendor's software suite? Will third-party applications be required to provide complete functionality?
- Will custom applications be required to provide the functionality the organization needs?
- How can the organization best use intranet and extranet technologies to disseminate information to users?
- How can CAD, GIS, and electronic document management systems best be integrated to an overall solution?
- How can metadata best be used to document the appropriate uses of the datasets?
- What improvements to computing infrastructure—servers, switches, routers, local area networks, wide area networks, and client personal computers—will be required to support the large GIS datasets?
- How can information be integrated from the various existing information systems to facilitate GIS access?
- What additional base mapping is needed as a foundation to support the utility GIS?
- What techniques are appropriate to consider for utility data conversion? A typical decision tree for evaluating utility conversion options is shown in Figure 3.3.
- What graphic and nongraphic database standards are best suited for the organization's needs?
- What tools and techniques can be used to automate any field data collection process?

The consultant must be able to answer these questions and develop order-of-magnitude cost estimates for each viable option before the visioning process can begin.

CONDUCT VISION WORKSHOP. At this juncture, all project team members meet for a one-day vision workshop. The purpose of this workshop is to lead team decisions that regard

- The applications that will be initially developed for the GIS,
- The technology that will be used to implement the GIS,
- The data that will need to be developed to support the GIS, and
- The staffing needs to support these applications.

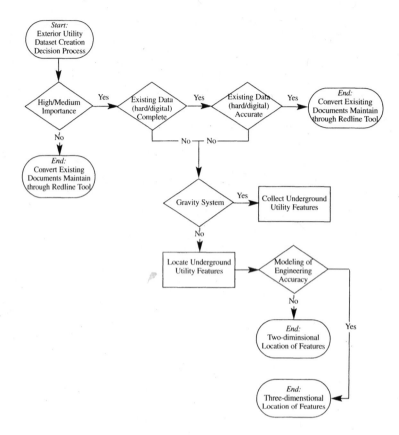

Figure 3.3   Data conversion and collection decision process.

A typical workshop can follow this pattern:

(1) Review the needs analysis summary prepared during the previous step.
(2) Describe the proposed applications and the data needed to support each of them. When new data requirements are extensive, quantify the relative costs. Also define the opportunities for business process improvement.
(3) Define other key issues (technology, staffing, organizational alignment, etc.) that must be resolved before the GIS can succeed. Present relative costs associated with each issue.
(4) Describe or demonstrate some practical examples of the use of GIS in other organizations. These presentations should typically be more specific to issues raised during the needs analysis than those presented during the project kickoff meeting.
(5) Use team decision-making techniques to prioritize the initial applications and corresponding data requirements of the GIS.

The results of the vision workshop will lay the foundation for the completion of the implementation plan.

Step 3: Develop the Implementation Plan. *DEFINE GEOGRAPHIC INFORMATION SYSTEMS DATA NEEDS.* Data conversion and creation is almost always the largest cost and time component of any GIS project. Because of these constraints, data collection needs should be evaluated for each application that the organization selects. These needs will be consolidated into an overall plan for acquiring or converting data needed to support the GIS.

DEVELOP SYSTEM ARCHITECTURE. At this stage, the consultant must evaluate potential technology solutions for identified GIS needs. The key components to be evaluated are the core GIS software and related products. This step produces specific recommendations for the hardware, software, networking, and integration that is required to implement an enterprise GIS.

DEVELOP IMPLEMENTATION STRATEGY. The consultant team must develop a detailed enterprise GIS implementation strategy. The strategy will include the following tasks:

- Phased purchasing of GIS and related software, hardware, networking, and peripherals.
- A phased plan for conversion and development of data. All GIS conversion efforts should include a pilot project to test the conversion approach.
- Definition of commercial-off-the-shelf and custom applications that will be purchased and developed.
- Detailed description of the processes affected by GIS and potential areas of improvement.
- Defined staffing and training needs, both for the organization's full-time GIS staff and other GIS users.
- Development of a widely accessible data catalog so that users know which data are available and their sources.
- Defined tasks that will likely require outsourced services.

PREPARE PRELIMINARY REPORT. The consultant must document the results of all steps to date and submit the results to the project team. This report must also include a detailed plan for the implementation of the GIS in stages, including the design stage, the pilot stage, and the full conversion and implementation stage. The description of each stage must include a detailed schedule, staffing needs, and costs.

REVIEW THE DRAFT REPORT. The project team will meet again to review the preliminary report and finalize plans for phased implementation of the GIS.

PREPARE FINAL REPORT. The consultant will then prepare a final report to document the project. Although, in most cases, the organization will probably continue to use the consultant for follow-on tasks, this plan must be of sufficient detail that the organization can implement the GIS without further assistance from the consultant.

PRESENT TO EXECUTIVE STAFF. In many cases, the results of the implementation planning process must be presented to the organization's executive staff before permission will be given to move onto the next stage.

Related Steps. A number of additional planning-related steps are required before primary data and application development tasks can be started. These tasks, while not typically considered part of the implementation planning process, are important and must be included either at the end of the planning process or at the beginning of a design process. These include development of the following:

- Database design,
- Data-conversion plan,
- Quality assurance/quality control plan,
- Data-maintenance plan, and
- Requirement specifications for each application.

THE IMPLEMENTATION PLAN DOCUMENT. The implementation plan document is the tangible product of the planning process. However, it must be stressed that the document is not, by itself, the key to the planning process, but rather the result of the process.

This example report outline includes the following sections:

(1) Introduction,
(2) Existing conditions,
(3) Needs assessment,
(4) Recommendations, and
(5) Implementation strategy, budget, and schedule.

Within sections three and four, parallel structures are used, with the organization based on the following categories:

- Data,
- Standards,
- Applications,
- Business processes,
- Technology, and
- Staffing and organization.

These categories are also carried forward into the budget and schedule section.

DOCUMENT OUTLINE. An outline of a typical implementation plan is shown. Each organization will have its own unique requirements that will

require variation from this outline. An example of a GIS implementation
outline is as follows:

(1) Introduction
  Document Organization
(2) Existing Conditions
  Organization
  Kickoff Meeting and Interviews
  Expectations and Needs
  Current IT Infrastructure
  Geographic Information System Support and Development
  Organizational Support for GIS
  Department Existing Conditions
    Department 1
      • Existing Data
      • Geographic Information System Needs
(3) Needs Assessment
  Introduction
  Data
    Data-Related Subsections
  Standards
    Standards-Related Subsections
    Metadata
  Applications
    Applications-Related Subsections
  Business Processes
    Business Process-Related Subsections
  Information Technology
    Desktop Personal Computers
    File Servers
    Networking
    Geographic Information System Software
  Enterprisewide Staffing and Organization
    Staffing and Organization Needs
  Summary of Needs
(4) Recommendations
  Introduction
  Data
    Recommendation #1, etc.
  Standards
    Recommendation #2, etc.
  Applications
    Recommendation #3, etc.
  Business Processes
    Recommendation #4, etc.
  Staffing and Organization
    Recommendation #5, etc.

INVESTIGATE THE PROCUREMENT OPTIONS

A large GIS procurement requires thorough consideration of the resources required for both implementation and maintenance of the developed GIS. The business case must include consideration of all issues, particularly data migration, if it is to truly represent all costs and benefits as well as set future measurement of a "break-even" point for success of the system. Organizations are typically faced with three development scenarios

- Build the GIS using in-house resources,
- Build the GIS using external resources, or
- Build the GIS using a combination of in-house and external resources.

The extent to which an organization uses external resources depends on what portions of the overall data conversion effort are to be conducted externally and what portions are to be conducted in-house.

Certain aspects of this decision process are straightforward as some resources are so specialized as to require an external consultant. For example, if a light image detection and ranging project is being conducted, the expertise and equipment will not be available in-house. Other services may be available within the organization. These can include planning services, conversion services, database design, and so on. True success requires a partnering approach between the organization and the selected vendor.

There are many external resources available, ranging from specialized GIS vendors to consultants who develop "turn key" GIS projects. Regardless of the capabilities required, the selection of the vendor is the first step in ensuring the success of the overall development. A number of factors should be considered when evaluating potential vendors, including

- Market presence,
- Technical capabilities,

- Company history,
- Geographic locations,
- Ownership,
- Project team experience,
- Project plan,
- Quality assurance/quality control approach, and
- Cost.

BUDGET AND PLAN FOR GEOGRAPHIC INFORMATION SYSTEMS

A geographic information system requires an initial investment and delivers a future return. Like any important investment, a GIS requires significant funding. Typically, funding is required for the following purposes:

(1) Computer hardware and networking infrastructure;
(2) Geographic information system and related software;
(3) Consultants for planning, design, and ongoing support;
(4) Data-conversion services;
(5) Custom software application development;
(6) Internal staff to maintain data and manage the GIS; and
(7) Ongoing maintenance costs.

In the past, GIS was treated as a technology-driven effort. Because of this, much of the funding for a GIS was directed at hardware, software, and computer networking. In some cases, GIS projects were the driving forces for local and wide area networks. Today, technology costs are much lower and, in most cases, organizations already have developed the technology infrastructure that is needed to support a GIS.

Figure 3.4 shows the typical breakdown for relative costs associated with initial GIS development for an organization. This chart shows that converting from paper to digital data is the largest portion of the costs associated with GIS development. This is similar to a construction project. Construction costs far exceed those of the planning and design stages.

Many utilities are reluctant to migrate to different or newer data formats because of the investment and effort that was required when they first implemented CAD or GIS. Because data are already stored in a digital format, costs for data migration are significantly lower than costs associated with initial data creation or automation.

In the sequence of GIS development, planning and design come first and therefore are generally affordable. However, they do not create any financial

Typical GIS Costs

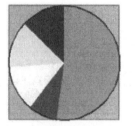 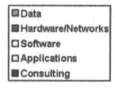

☐ Data
■ Hardware/Networks
☐ Software
☐ Applications
■ Consulting

Figure 3.4   Typical GIS costs.

return so they are just the beginning of the investment. Additional investment is required through the procurement of hardware, software, and data before the GIS begins to generate benefits, often two or three years after the initial investment.

FUNDING STRATEGIES. Different organizations take different approaches to acquire the funding needed to develop a GIS. In addition to funding from capital and operating budgets, potential funding sources include

- Grants,
- Other agencies, and
- Other funding sources.

Most organizations use a combination of sources. Each source has a different set of issues.

Funding through Grants. Grant funding for GIS development is generally hard to find in the utility arena. Enough funding to start small projects might be available for small communities from state sources.

Some larger software vendors have developed grant programs. These programs provide software to assist with beginning GIS efforts. This "free" software can be a mixed blessing. It can come with annual maintenance costs that run as high as 25% of the purchase price. Some organizations pay more over time for maintaining the "free" software than the value the software provides because the software often cannot be used or the organization may not need it.

Other grants sources may include

- State agencies (for example, environmental or finance departments) and
- Federal agencies (for example, U.S. Environmental Protection Agency, U.S. Geological Survey, or U.S. Army Corps of Engineers).

Funding through Other Agencies. Utility GIS projects typically are begun as a part of an organizationwide or multiorganization effort. This can be a great help to offset some of the costs of a GIS because the costs for common datasets can be shared by all participants. Although this is not really an alternate funding source, sharing these costs can be important because it reduces the need for funding. True assistance from other agencies in funding utilities is rare.

Other Funding Sources. Some organizations operate a GIS as a utility. In this arrangement, the GIS program charges other departments in the organization for its services. Assuming that costs are allocated based on services provided, this approach has an advantage. It links the costs and benefits more fairly than other methods. However, this method does need to be kept simple— otherwise, the organization spends precious time documenting and chasing internal costs at the expense of serving customers.

COST JUSTIFICATION. One of the biggest barriers to developing GIS is justifying the costs. It is relatively easy to quantitatively define the costs needed to develop a GIS. However, it is much more difficult to quantitatively define the benefits.

Most utility managers understand the value of GIS and want to implement it but need to quantify benefits to compete for funding with schools, programs for senior citizens, and the like. Often they are faced with the daunting prospect of going before their city council or agency board of directors to request hundreds of thousands of dollars for a GIS at the same time their governing body is faced with budget cuts. Therefore, cost justification is often a critical step for a GIS project

As discussed in Chapter 2 of this book, there are many benefits of a GIS. Most of the benefits have a direct financial benefit to the organization. However, it is difficult to generate a meaningful value to each benefit. This is particularly true when GIS is used to transform processes such that entire steps within an organization's processes are eliminated.

A Dayton, Ohio, based GIS consulting firm conducted a study of a number of public and private utilities (Woolpert, 1999). These utilities were surveyed regarding how they developed quantitative GIS benefit amounts, either before starting GIS projects or after the GIS was operational. Of the 13 utilities that were surveyed, none attempted to formally define the benefits before the GIS project was begun. Only two had attempted to define the benefits once the GIS was operational.

However, as demonstrated in the consulting firm's survey, some organizations have chosen to try to quantitatively define the benefits. Some of the specific benefits for which quantitative estimates have been derived include

- Time savings that field crews experienced because they had better routing information;
- Time savings field crews experienced because they had better infrastructure location information;

- Reduced time in map maintenance (although sometimes this time increases);
- Reduced risk when responding to events where uncertainty is common in basic information, for example, storm intensities or frequencies;
- Reduced time in field inspections (although sometimes this time increases);
- Better quality designs because of better information, leading to fewer field changes;
- Better inspections, which result in better quality construction projects;
- Reduced time in preparing hydraulic models;
- Reduced time in maintaining infrastructure cost accounting information;
- Savings from using GIS to better plan multiple capital projects; and
- Savings from using GIS data to determine which capital projects to undertake.

In some cases, organizations have developed formal cost–benefit analyses and used them to determine whether GIS projects should be developed. Unfortunately, there are many unknowns in this process. Often, a cost–benefit analysis is an exercise with a predefined outcome (for or against the project) that is based on management's beliefs.

In most organizations, GIS is justified because of the clear benefits that it can bring—even if those benefits cannot always be quantitatively defined. A geographic information system is such an improvement over any other approach that it is virtually a requirement to effectively manage utility systems today—like many other technical advancements that we take for granted today such as telephones, personal computers, and e-mail.

DATA AS AN ASSET. When organizations do invest in GIS, they also need to understand that they need to make a commitment to maintain the information. The value of this information should be considered as an asset to the organization, and the information must be maintained to maintain the value. Some organizations have gone so far as to consider GIS data collectively as a capital asset that can be depreciated over time.

CONCLUSION. In operation and GIS environments today, strategic planning is an evolutionary process. However, many organizations have fixed their strategies so that they are unchanged and unexamined for five- to ten-year timeframes. To develop a successful GIS development strategy, it is critical to adhere to the essence of strategic planning—flexibility must be built in to the process.

Organizations today focus on the short term—whether that is good or bad, GIS developers must respond to this environment. Great GIS plans mean nothing to many organizations if there is nothing to show for them today. While we should not necessarily short-change long-term plans for a GIS, strategic implementation for technology often involves investing resources to provide short-term benefits. The bottom line is practicality.

REFERENCES

Obermeyer, N. J.; Pinto, J. K. (1994) *Managing Geographic Information Systems;* The Guilford Press: New York; pp 128–147.

Woolpert LLP (1999) Louisville Water Company AM/FM/GIS Implementation Plan; Louisville Water Company: Louisville, Kentucky; pp 8–1 to 8–16.

SUGGESTED READINGS

Somers, R. (1994) GIS Development Alternatives: 10 Years vs. 10 Days. *Geo Info Systems*, **4** (9), 2–6.

GIS Strategies & Issues—Strategies for Managing and Utilizing GIS in Your Organization. Retrieved. http://lagic.lsu.edu/gisprimer/strategies.asp (accessed Feb 23, 2002); hosted by Louisiana Geographic Information Systems Council, Baton Rouge, Louisiana.

Chapter 4
Outline: Data Conversion and Database Development

*I*NTRODUCTION

Geographic information systems (GIS) applications for water, wastewater, and stormwater utilities support a wide range of planning, design, operations, and maintenance activities. The benefits of an operational GIS system to a utility's operations are well-documented throughout the GIS industry. *The applications that can be carried out with a functional GIS system are directly related to the quality, content, currency, precision, and positional accuracy of the data.*

As GIS implementations mature, more and more organizations that have developed their planimetric–topographic and parcel base maps are now in the planning or implementation stages of a utility conversion project. Many organizations that had previously converted utility information are also now looking to enhance the quality of their utility database through the conversion of additional features, enhancement of the positional accuracy, or through integration of their GIS data with modeling or customer information systems.

For those users in the beginning stages of their GIS implementation effort, the database development phase is often the most expensive and time-consuming part of their implementation. Other users who have completed initial conversion activities often find that the database enhancement process and maintenance effort is an ongoing activity in which new technologies allow for further expansion and enhancement of their database. Early investments into increased planning at the project planning and design stage will be less expensive than the cost of modifying and upgrading a poorly conceived database.

This chapter discusses basic concepts and key issues associated with the creation of a GIS database to support utility operations and applications. It provides an overview of the data conversion process, techniques for database development, and recommendations for effective management of a data conversion project.

Often abbreviated as GIGO, "garbage in, garbage out" is a famous computer axiom meaning that if invalid data is entered into a system, the resulting output will also be invalid. Although originally applied to computer software, the axiom holds true for all systems, including decision-making systems such as GIS systems.

This concept is critical to understanding what makes a successful data conversion project. If the quality of the data to be converted is suspect, steps need to be taken before initiating the data conversion process, or specifically built into the process, to enhance the quality of the sources to better prepare them for conversion. Too often, conversion projects are treated as opportunities to clean up or normalize inconsistencies in records management that have occurred over decades, and in some cities over hundreds of years. Even worse, for some projects data is simply converted without a clear understanding of the quality, consistency, or accuracy of the information to be converted.

CONVERSION PROCESS

A data conversion project should follow a structured approach to ensure consistency, integrity, completeness, and overall accuracy of the database. Figure 4.1 depicts the conversion process. Major steps are described below.

Requirements definition. This task involves defining application and data requirements. Requirements for graphic and nongraphic database content, accuracy, scale, and data structure should be identified.

Source evaluation. Involves examining and cataloging source materials. Sources should be examined for accuracy, completeness, and interpretability. As part of this review, a determination of scrubbing or preconversion data preparation requirements should be made.

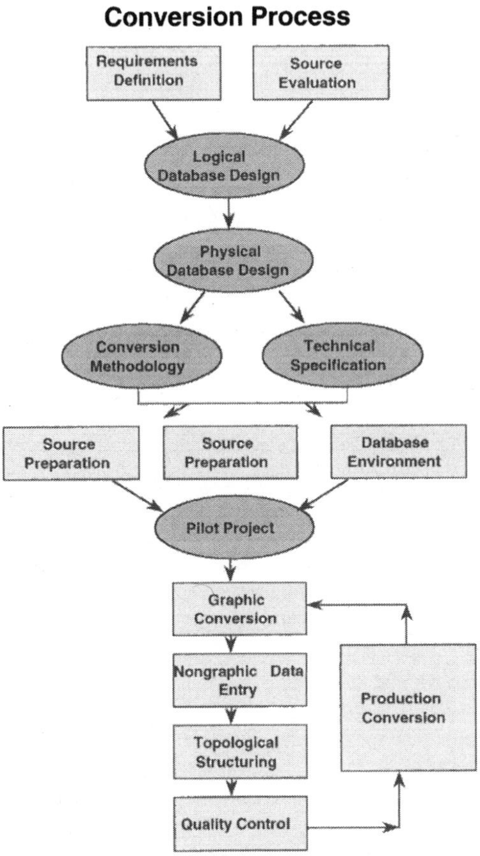

Conversion Process

Figure 4.1   Conversion process.

Database design. Typically a two-phase process, involving first the development of a logical database design, and then development of a physical database design. The logical design process involves defining the following:

- Features to be captured,
- Attribute data requirements,
- Annotation and text placement rules,
- Data format requirements,
- Placement rules and guidelines, and
- Object relationships between entities.

The database design should also include some considerations as to the applications that are envisioned and what their data requirements will be.

Physical database design. The physical design is a GIS software or database specific design that reflects the data structure or format in which the data will reside (Shape file, Arc Info coverage, Oracle Spatial, Spatial Database Engine (SDE), DGN (design file), DWG (drawing file), Mdb, Smallworld, etc.). Major components include the following:

- Database structure and layer geometry;
- Specific features to be included on each layer, and the schema or rules to be used for encoding intelligence (e.g., directionality and spatial relationships) into each feature;
- File size, structure, and naming conventions;
- Attribute table names, format, schema, and key identifiers to provide links to attribute data sources. Also, identification of required attribute fields;
- Network structure and connectivity rules; and
- Symbology standards for cartographic or plotting purposes.

Technical specifications. As part of the design effort, technical specifications that detail the approach sequence, product, accuracy standards, and the production schedule should be developed.

Conversion methodology. The conversion methodology should be based on the database design to be implemented, source materials to be used, and available budget. These items should be the "driver" for the conversion methodology.

Database environment. The database environment needs to be established. This involves creating a directory and file structure or, in the case of a relational database structure, assigning disk resources upon installation and configuration, into which the converted data will be inserted. It typically also involves establishing access privileges and user accounts, creating symbol libraries, and plotting applications.

Source material preparation. Depending on the project, source material preparation or scrubbing activities may entail a significant level of effort. At a minimum, a set of source materials should be assembled and reproduced before conversion.

Data conversion application. For any large conversion effort, a production conversion application should be developed. Development of a conversion application involves creating menus and writing macros, programs, and other automated routines to streamline the conversion process and minimize operator error. It also includes the development of automated quality assurance/quality control (QA/QC) routines. Select personnel should be assigned to the application to update it whenever changes occur to the database design. When the changes are implemented, the application should reflect the changes to enforce all new parameters of the project.

Pilot project. An evaluation of the design, conversion methodology, and technical specifications should be conducted by developing data for a two- to four-sheet pilot area before proceeding to the production phase of a conversion project.

Graphic entry. The entry of graphic data should be accomplished using a structured conversion application that ensures that data is entered into the database in a consistent and logical format that is in accordance with the design.

Attribute entry. Attribute data entry may be closely integrated with the graphic data entry process or it may be a separate step in the process. As with the graphic data entry process, nongraphic data should be entered using customized data entry screens and forms. Nongraphic entry may also involve loading data from existing personal databases, spreadsheets, or corporate relational database management systems.

Topological or rule-based data structuring. Creating a network data structure is an integral step in the conversion process. It may be integrated to the production conversion process or done as a separate step after the graphic and attribute data is converted. Ideally, it should be performed as an integrated part of the conversion process. The building of topology involves the use of automated routines to define spatial relationships such as connectivity and adjacency in the database. For sewer or stormwater information, it may also involve flow directionality.

Quality control. Quality control checks should be incorporated to several steps of the production process. Checks for data completeness, content, symbology, data format, and adherence to design should be conducted. A comprehensive QC program that involves the use of automated and manual checks must be implemented for any large-scale conversion effort. Quality control should also include field checks to verify positional accuracy.

Outline: Data Conversion and Database Development   *85*

Production conversion. After the pilot project is completed, the procedures, specifications, and application programs designed during the pilot project should be modified to streamline the production process.

In addition to the conversion steps defined, metadata information needs to be established that encapsulates elements of all items defined above. This is essential to communicate the proper use of the data that is housed.

DEVELOPING A BASE MAP TO SUPPORT UTILITIES OPERATIONS

BASE MAP DEVELOPMENT. The base map serves as the spatial foundation upon which utility features are referenced. The base map may be developed by a utility, public works, or engineering department, or it may be acquired from another department or organization. Decisions regarding base map development are critical because the base map serves as the reference layer for the utility information in the database. The base map is a critical determinant of the overall spatial accuracy of the database and directly influences the scale at which the information can be used.

Several alternative base map products are used for utility operations. The different types of base maps are briefly described below.

Digital Orthophotos. A digital orthophoto is a photogrammetric base map that is created by obtaining digital imagery from a large format digital camera or scanning aerial film produced from analog aerial photography using a precision image scanner. Figure 4.2 depicts an example of a digital orthophoto.

The source materials used for digital orthophotos are aerial photographs, surveyed ground control data, camera and scanner calibration data, and a digital elevation model. A digital orthophoto provides a raster (photographic image) land-base product. All features visible on the photography appear as images in the database. This image may be displayed in the GIS as a base map for other vector layers.

Typically, digital orthoimage base maps are acquired at levels of resolution ranging from 0.25" pixel size to 2" pixel resolution. This level of resolution is often produced for map products designed for display at 1" = 40" scale to 1" = 200" scale. Orthoimage products may be custom developed as part of a photogrammetric project or acquired from a firm that sells commercial imagery. Orthoimagery may also be obtained from several satellite imagery providers.

Accuracy standards and coordinate datums used for orthoimagery are consistent with those used for other photogrammetric map products. All imagery will also need to be appropriately georeferenced. Georeferencing

Figure 4.2   Example of digital orthophoto.

enables the imagery to be placed in its precise coordinate location so that the imagery can serve as a backdrop to other spatial data sets.

Planimetric Map. A planimetric line map is a photogrammetric base map that is created by compiling (digitizing) map features from aerial photography using photogrammetric workstations. Figure 4.3 depicts a picture of a planimetric map. A photogrammetrist interprets and digitizes selected features that are visible on the aerial photography. The key issue associated with development of a planimetric base is the scale and accuracy required and the feature content. Primary features generally required include buildings, pavement edges, and major hydrography. Secondary features include railroads, parking lots, driveways, piers, vegetation, and open space. Other features include street and stream centerlines, wall structures, utility points (i.e., traffic signals and street lights), fire hydrants, manhole covers, and contours (index and intermediate). In a utility conversion project, it is important to determine if infrastructure features are to be mapped from aerial photography or converted by other techniques.

Parcel Base Map. Although a photogrammetric landbase is more commonly used as a base map for utility operations, a cadastral base is used by some water and sewer utility organizations as a base to reference utility information.

Figure 4.3   Example of planimetric map.

A cadastral base contains ownership information such as parcel and lot boundaries, rights-of-way (ROWs), and subdivision boundaries, among other items.

*U*TILITY *CONVERSION* *TECHNIQUES*

Many different techniques can be used for converting facilities data. Factors that are taken into account when deciding on the appropriate conversion technique include

- Database design and application requirements,
- Quality and availability of source materials,
- Accuracy requirements, and
- Funding availability.

The primary techniques used by utility organizations for graphic data conversion are described below. The specific technique to be used for a project should be defined as part of the specification development process. Figure 4.4 illustrates facility conversion techniques.

DIMENSIONAL ENTRY AND CONSTRUCTION. Dimensional entry, or offset entry techniques, involves the use of dimensional information to place

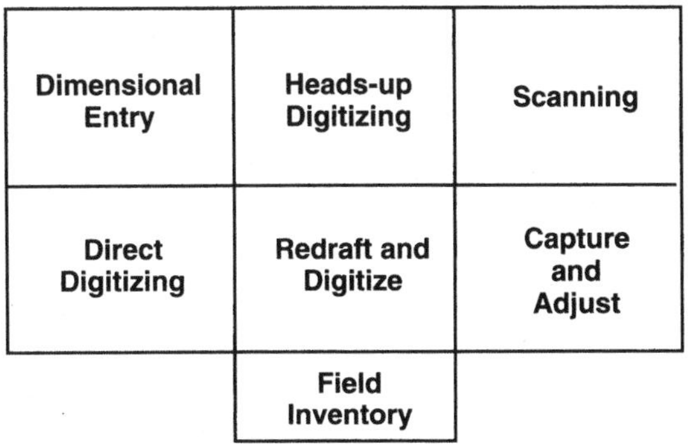

Dimensional Entry	Heads-up Digitizing	Scanning
Direct Digitizing	Redraft and Digitize	Capture and Adjust
	Field Inventory	

Figure 4.4   Facility conversion techniques.

line (i.e., mains and laterals) or point (i.e., manholes, hydrants, and catch-basins) features. Typically, dimensional information is contained on "as-built drawings" only and is often not shown on an agency's index maps. Use of this technique may therefore involve scrubbing existing maps to indicate offset distance from base map information (e.g., curb line) in the database. After the distance is determined, the automated drafting capabilities of a GIS are used to precisely place a feature at the exact offset specified. Features are typically placed at a specified offset from the curbline or the ROW line. For example, if a water line is 10.8' from the ROW, a ruled line can be drawn precisely 10.8' from the ROW.

This technique provides an opportunity to achieve a high degree of relative accuracy. The major disadvantage associated with the use of this procedure is the lack of offset information contained on existing maps. The other major problem is that the ROW or curb line information may not be very accurate, or may have changed since the as-built was initially recorded. Figure 4.5 depicts an example of a dimensional entry and construction technique.

DIRECT DIGITIZING. Direct digitizing techniques involve mounting a source map on a digitizing tablet, and through the use of a digitizing puck, registering the source map to three to four control points and directly digitizing facility information. Direct digitizing techniques involve no fitting or adjust-ment of the source to the reference base on a localized basis. All of the registration work is accomplished through the use of registering maps to control points. Typically, this technique is only used when the existing maps are being replicated in the GIS database, or if the source maps (the digital base map and the facility map source) are of a consistent accuracy. If the accuracy of the existing facility map is different than that of the land base that has been developed in the GIS, this technique is typically not appropriate. Figure 4.6 depicts an example of a direct digitizing technique.

HEADS-UP DIGITIZING. This technique involves an operator digitizing features on-screen in a heads-up mode at a workstation. No mounting or calibrating to control points at a digitizing tablet is required as part of this

---

Outline: Data Conversion and Database Development   **89**

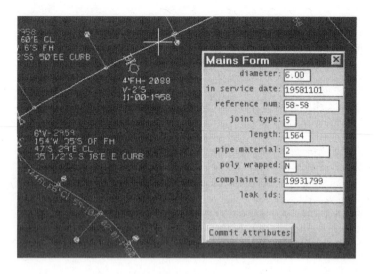

Figure 4.5   Example of dimensional entry and construction technique.

process. The source maps are used as a reference base for placing facility information. No direct digitizing of the source occurs using this approach. The use of this technique essentially involves recompiling features digitally on screen. Appropriate reference features must exist in the drawing to reference the facility information.

Heads-up digitizing techniques are often used in combination with scanning techniques. For example, as-builts or index maps may be scanned and georeferenced to base map features in a GIS database. Further refinement of

Figure 4.6   Example of direct digitizing technique.

the georeferencing may be done on a block-by-block basis. After the source file (scanned image) is positioned correctly, features may be digitized directly with the scanned image being used as a backdrop.

Scanning techniques can be used only if the sources are of a good overall quality and consistency. If scanning techniques are used in combination with heads-up digitizing, calibration must first be established to ensure that scanning "slips" or scale distortions do not occur.

REDRAFT/DIGITIZE. In those cases where a new and accurate land base has been acquired and the existing water facilities maps are very inaccurate, or if the source to be used is "as-builts," it may be necessary to manually redraft facility information onto the new base map before digitizing.

After the maps have been redrafted, direct digitizing techniques can then be used to enter the data into the database. The redrafting process ensures that the correct spatial relationships between features are defined before data capture. The redrafting process typically only involves a redrafting of lines and symbols; text information is entered as a separate step in the process and does not need to be redrafted. Different colors are used during the redrafting process to aid the digitizer in differentiating features.

CAPTURE AND READJUST. A common method used for facility conversion involves capturing the data digitally either through scanning or direct digitizing, and then readjusting the data to the base map. This technique is often necessary because of the differences in relative accuracy of the source maps being used and the GIS base map. The adjustment approach to the land base may be on a sheetwide, block-by-block, or street- or facility-segment basis.

FIELD INVENTORY AND INPUT. Under this approach, surveyed data is obtained for point features to be included in the database. Specific techniques for field data acquisition are described below. When a point file is created, linear features are converted by "connecting the dots" using an appropriate index map as reference. Annotation and attribute information would be captured from the index source maps.

*F*IELD *DATA ACQUISITION*

At an ever-increasing rate, many organizations are looking to capture their street infrastructure features via field surveying techniques involving global positioning system (GPS) technology. These techniques provide the opportunity to significantly increase the level of positional accuracy of the utility infrastructure. At the same time, they also provide an opportunity to collect additional information in the field at the time of capture. These techniques vary extensively in terms of accuracy, productivity, cost, complexity, and practicality.

The environment in which they are applied (i.e., urban versus rural, heavily vegetated versus open space) also influences whether or not each of the techniques will be successful. Table 4.1 summarizes the most common techniques

Table 4.1 Different techniques of field data acquisition.*

Surveying technique	Expected horizontal accuracy (m)	Expected vertical accuracy (m)	Accuracy assessment	Productivity	Reliability	Safety/ logistics	Completion percent	Estimated relative cost	Notes
Conventional surveying: total station/traversing/leveling	0.05	0.05	●	○	●	○	●	○	May include static GPS stations as base control.
Field measurement: tape, compass, wheel	1.2	0.06	○	○	❖	○	●	○	Uses identifiable base map features as control. Accuracy dependent on base map accuracy.
GPS: differential GPS	1.0	2.0	❖	❖	❖	❖	❖	❖	Requires visibility to minimum of 4 GPS satellites. Assumes post processing/occupation of features <1 minute.
GPS: Real-time kinematic	0.03	0.03	●	○	●	○	❖	❖	Requires visibility to minimum of 4 GPS satellites. Assumes post processing of features. Accuracy may vary depending on postprocessing procedures used.
Photogrammetric: stereocompilation	0.06	0.02	❖	❖	❖	●	○	❖	Compilation dependent of scale of photography. Photography smaller than 1" = 600' not adequate for most utilities.
Reconstruction: from ROW or curb	1.20	0.06	❖	❖	❖	❖	❖	❖	Accuracy dependent on base map accuracy. Technique uses base map features as a guide to determine the placement of utility features. No surveying techniques are used with this procedure.
GPS/INS: with feature extraction—Van based	0.03–0.5	0.03	❖	❖	❖	●	❖	❖	Features extracted after georeferenced data collected.

*GPS = global positioning system; INS = inertial navigation systems; and ● = high, ❖ = medium, and ○ = low.

and their approximate accuracy. Typical water, sewer, and stormwater utility features that could be collected in the field include

- Manholes,
- Catch basins,
- Fire hydrants,
- Water valves,
- Pumping stations, and
- Stormwater outfalls.

All of the field surveying techniques have the advantage over photogrammetric capture techniques in that they allow for positive feature identification in the field (e.g., sewer versus stormwater manhole).

The conventional surveying techniques involving the use of total station traverses can be very accurate; however, in highly urbanized areas they are often impractical given the issues of traffic and safety management concerns. Costs can also be significant.

Positional accuracy requirements greater than submeter will demand the use of real-time kinematic (RTK) GPS techniques because differential methods (either real-time or post-processed) cannot reliably achieve this level of accuracy. Many organizations specify $+/-1'$ accuracy so that features are of a greater accuracy than their landbase. This necessitates the use of RTK techniques or very large scale photogrammetric mapping techniques (e.g., photo scales of $1' = 300'$ to $1' = 400'$). In general, the vertical accuracy of RTK GPS is three times worse than the horizontal accuracy. It should be noted that the accuracies stated above require additional field calibration techniques.

In a relative sense, photogrammetry, reconstruction, and GPS–inertial navigation systems (INS) are more cost-effective for larger geographic areas than GPS or conventional survey methods only (for capture of the same number of features).

Pedestrian techniques using GPS methods can achieve high levels of reliability in terms of feature identification and correct attribution. The capture rates may be moderate (75%) to low (50%), depending on the time of year (leaf-on or leaf-off conditions) and the characteristics of the project area, as determined by objects such as buildings, which obscure the GPS satellite signals.

Many organizations look to achieve over 95% capture in the field. If this is the case, a hybrid of techniques must be used. For example, to achieve 100% of features to an accuracy of $+/-1'$, the RTK–GPS surveys must be supplemented by conventional surveying methods such as total station traverses.

If field capture rates approach 100%, there is often a situation of diminishing returns, whereby considerable effort, expended at great cost, is required to survey the final few percent of the features. This can lead to scheduling problems, budgetary issues, and dissatisfaction on the part of both the client and the contractor. The 100% capture for utilities from the field is also not an appropriate goal because of buried, obstructed, or highly inaccessible features.

Any procedure used should incorporate methods and procedures for supplementing data.

One of the newer technologies beginning to be used for utility capture is the use of a specially equipped vehicle with an on-board GPS–INS and an array of high resolution cameras to capture imagery which can be post-processed and georeferenced.

Use of this technology involves driving the roads within an area to collect digital imagery and GPS-based positioning information. After this data is postprocessed, features are collected in three-dimensional form on screen. The systems are used to collect features that are located on roads, along road ROWs, and in parking areas.

These mobile mapping systems may offer, depending on the project, some advantages over traditional GPS surveying techniques, including increased safety, shorter schedules, increased efficiency, higher accuracy, and robustness in urban environments. Another additional advantage is that referenced imagery can be used for photologging-type applications and may minimize the need to go out into the field to investigate a site. Figure 4.7 depicts an example of a mobile mapping system (GPS) and Figure 4.8 depicts the equipment used.

Figure 4.7   Example of a mobile mapping system.

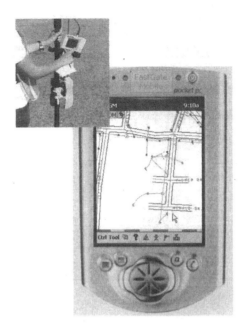

Figure 4.8   Field inventory tools.

DATABASE CONTENT

As part of the facilities conversion project, determining the content to be included in the utility database is an important consideration. Most utilities have different feature requirements, depending on their applications. Table 4.2 depicts an example of content of a database. Content requirements go hand-in-hand with the sources to be used and the method to be used for conversion.

All of the individual entities in the system should be uniquely feature coded so that individual features within a system (i.e., water, sewer, stormwater) can be logically distinguishable from one another. For GIS systems that use an object-oriented data structure, this feature coding is often incorporated to the entity name, and for layer-based systems it is typically stored as an attribute (either numeric or character) of each element. Examples of these features included the following:

(1) Manhole Cover = MANCOV (Entity Name) and
(2) Manhole Cover = 100; Fire Hydrant = 230 (Numeric or Character attribute).

Another key consideration for content is to determine which features include required to support modeling. This is particularly important in defining attribute requirements.

Table 4.2   Example of database content.

Water facilities	Sewer facilities	Stormwater facilities
Pipes	Pipes	Pipes
Water main pipe	Sewer pipes	Stormwater pipes
Fire pipe	Gravity main	Combined pipes
Abandoned pipe	Force main	Culverts
Valves	Combined pipes	Stand pipes
Valves (division, butterfly, check, blowoff, pressure reducing, air relief)	Abandoned pipes	
Pressure-reducing valve	Intercepter sewers	Manhole/point
Valve chamber		Storm manholes
Hydrants	Manhole/point	Catch basins
Hydrant tee	Manholes	Inlets
Hydrant valve	Lamphole	End sections
Hydrant branch	Endwall	Outfalls
Hydrant	Sewer overflow	Siphons
Fittings		Wells
Tee	Services	Boundaries
Cross	Service lateral	Detention ponds
Reducer		Watershed boundaries
Coupler	Facilities	Infiltration basins
Services	Treatment plants	Hydrography
Service pipe	Pumping stations	Hydrography
Curbstop	Flow monitor stations	Paved ditches (wet/dry)
Corporation stop	Rain gauge stations	Unpaved ditches
Meter	Boundaries	(wet/dry)
Other features	Boundaries	
Pitometer tap	Sewer service area	
Isolation flange	Minibasin boundaries	
Anode/test stations		
Air vent	Facilities—water/sewer	
Vault	Raw/treated water reservoirs	
Air cock		
Pipe tap		
Water wells		
Pressure-monitoring stations		
Sampling stations		
Boundaries		
Water pressure zones		
Service area		

*D*ATA INTEGRATION

Most utility organizations have in-place operational work order management
and customer billing systems. Automated systems for meter reading or facilities
management may also be in place. Permitting systems and real estate files

(assessment databases) are also frequently accessed, and in some cases specific fields are maintained by the utilities department. Documents such as contract plans, as-builts, valve cards, and other paper documents may also have been scanned and stored in the document management system. Figure 4.9 depicts an example of data integration. As part of the data conversion effort, requirements for integration of these existing systems must be addressed.

Data that is maintained in each of these systems may be stored on mainframe computers in propriety databases, in relational database management systems using client-server or Web-based architectures, on PC-based databases, and on spreadsheets. A fully integrated environment involves integration of system and networking protocols and application integration.

Data that is converted must take into account requirements for integration with these existing systems. To facilitate integration, there must be relationships between the attribute databases in each system to be integrated. For example, a facility identification (ID) in an inventory database must match to the ID of a feature in the GIS. A contract number for a pipe segment must match to a contract number in an index to access images stored in a document management system. A customer account must match to a parcel or at least a street segment in the GIS by address. The integration may occur as part of the conversion process or as a separate process after conversion is complete. A custom application should be designed to support this activity. Its function, based on the current database design and source materials, will run multiple queries (tests) that will associate IDs with their corresponding features. An error report should also be included in the application to check the accuracy of the ID-to-feature association.

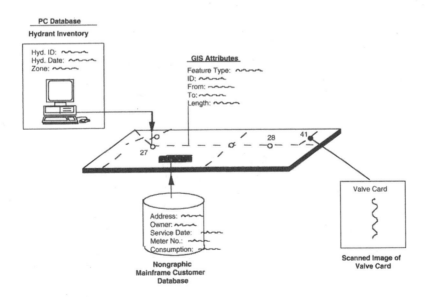

Figure 4.9 Example of data integration.

There are four common relationships that should be taken into account as part of a conversion process because they affect long-term integration opportunities.

- Facility records to facility features. Many inventories were developed before map-based facility identification systems were implemented. Even if a map-based facility ID system is in place, the map grid upon which the IDs are based may have changed. The facility identifiers (e.g., valve ID and manhole number) that are in use in existing inventories must be entered as attributes of the corresponding feature in the GIS. Correlating the database records to the correct feature locations can be a cumbersome process. If existing feature IDs are contained in an automated database, a "pick table" can be established to be used as the basis for associating IDs to a feature.
- Customer addresses to parcel IDs. To analyze customer or billing systems information geographically, there needs to be a relationship between both of these items. Typically, customer billing systems do not contain a parcel number, and assessment databases do not contain a site address that corresponds to the address in the billing system. The relationship of a property or building to address is often a one-to-many relationship. Therefore, a link between the customer service address and the parcel number needs to be developed. The typical approach is to try to match the premise address in the property appraiser's parcel file to the customer service address. Because the premise address is often not as accurate or as well-maintained as the billing address, the matching process may require a good deal of QC and address validation. One or both sets of addresses may also need to be reformatted or parsed to support the match.
- Customer addresses to site address. For some utilities, the address-to-parcel relationship is specifically developed as part of the GIS development process. In those instances, addresses are assigned to building polygons or to a point representing the address. This relationship, once established, will also facilitate integration of billing and landbase data.
- Facility features to customer locations. One of the most difficult relationships to establish defines the actual facility, which serves a particular customer, such as the water main or the sewer pipe to which the service is actually connected. With this connection established, applications can determine customers affected by system problems or construction or maintenance activities. Demand and estimated flow can be more precisely calculated. Based on the assumption that the closest pipe in the street serves each customer, creating buffers on pipes and intersecting with parcel polygons can make a partially automated match. However, situations such as multiple pipes in the street, corner lots, and rear services affect the reliability of this technique.

QUALITY ASSURANCE AND QUALITY CONTROL

Data developed as part of a data conversion project should be subjected to a series of comprehensive QC checks. Although the goal for any production process is to produce 100% compliant data, most organizations recognize that they must allow for a small percentage of noncompliant data.

QUALITY ASSURANCE AND QUALITY CONTROL (VALIDATION). Quality assurances, by definition, are those processes and procedures that are incorporated to the collection and production phase of a project to ensure that high-quality products are produced. These include such items as detailed specifications, attention to workflow processes, automation, checklists, menu pick lists, and design standards. Each step in the production process should be designed with quality in mind. Possible errors should be anticipated, and procedures are instituted to prevent errors from happening. Continuous review of procedures and corrective action to the processes will prevent future errors.

Quality control or validations are those measures implemented to identify errors after a production phase is completed. These include such items as automated geometry and attribute checks, design validation reports, and manual feature verification. Specific attribute values or a range of acceptable values can be verified. Relationships between different attributes can be checked. Figure 4.10 depicts an example of a quality control tool.

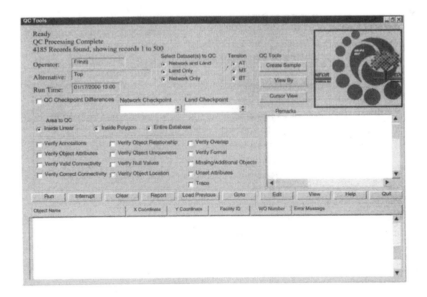

Figure 4.10   Example of a quality control tool.

Software for validation must be developed or acquired. Standardized screen displays can also be designed, and check plots can be customized so that errors are more easily identifiable. While manual or onscreen feature checking is time consuming, these types of inspections may be the only way to check some types of graphic features against source documents.

Validation activities should be considered like any other production process. The standards and steps should be defined. Responsibilities for each validation step should be assigned. Acceptance criteria should be standardized and realistic and based on how the data will be used.

Production schedules should be established to account for specific QC steps, and time for any corrective actions to occur. After any corrections are made, additional QC checks should be made to ensure the corrections were made properly.

Problem Resolution. Often, a validation process will uncover types of errors that stem from conflicting source documents. These are sometimes called *discrepancies* and will require research (sometimes called *problem resolution*). It is important to establish early who will perform the research to solve specific types of discrepancies. The solution may likely involve going to a secondary or, in some cases, a tertiary source to resolve the discrepancy.

Time limits for problem resolution should be established (e.g., 24 to 72 hours) so that production procedures are not adversely affected. Attributes can be used to flag features for which further investigation is needed to verify it correct.

GENERAL RULES FOR PRODUCING QUALITY. Some general rules about QA and QC include the following:

- Quality is the result of careful planning.
- Define the specifications for measuring data quality and the specific criteria for final data acceptance.
- Define the procedures for meeting quality standards. Procedures should be as standardized as possible. Validation processes must be documented.
- Errors should be detected as early in the process as possible. The later a mistake is found, the more costly it is to correct.
- All errors should be logged.
- Staff should be encouraged to report errors, and especially to recommend improved procedures to prevent errors. Changes to procedures should be tested outside of the standard process until proven valid.
- Acceptance criteria should be achievable and realistic. A lot of quality criteria are subject to interpretation. Define the criteria by the way that the data will be used.
- Ideally, all errors should be corrected, even if they are within acceptance limits.
- Assign responsibility for performing validation tests, researching discrepancies, and fixing mistakes.

UTILITY NETWORK DATA ACCEPTANCE CRITERIA. Each organization should establish, as part of the conversion program, the standards under which the data should be produced. Listed below are typical standards used to establish acceptance criteria. These standards may be modified depending on the specific requirements of any project.

Facility Network Data Acceptance Standards

• Deliverable Format/Design Adherence	100%
• Positional Accuracy (based on American Society of Photogrammetry or National Map Accuracy Standards)	90%
• Graphic Quality	98%
• Annotation Placement	90%
• Completeness	99%
• Priority Attribute Accuracy	98%
• Nonpriority Attribute Accuracy	96%

As part of establishing the standards, the unit area to be measured, and the rules for counting features should also be defined.

When the QA/QC procedures are established, a database should be designed to enable the QC personnel to not only record the types of errors, but also quantify the number of errors per type. Through implementing this database, reports can be generated that aid the QC personnel in verifying that the acceptance percentages are being met. If the percentages are not being met, QC personnel can communicate the discrepancies to the production department in a timely and precise manner.

DATA CONVERSION MANAGEMENT

Utility data conversion projects are typically large and complex undertakings. Successful project management requires dedication of personnel who know and understand utility records and data and have the tools to properly manage what is often a complex and multifaceted project.

Described below are a number of issues that should be addressed as part of the data conversion management process. Incorporating these activities to the project management process will increase the likelihood of success.

PROJECT DOCUMENTATION. It is important to prepare appropriate project documentation. A common mistake is to shortchange the documentation effort and to rely on verbal understandings (or misunderstandings). Key documents to be developed and maintained include the following:

• Technical specifications. This document should define the overall standards, features to be captured, accuracy requirements, sources to be used (quantities and characteristics), project limits, and deliverable products.

- Conversion workplan. The conversion vendor typically prepares this document. It should describe the specific internal procedures that will be followed in completing the conversion process. It should describe all activities related to the project workflow, including project setup, data preparation, production conversion, attribute entry, data translation, QC, and final data delivery.
- Database design. This document may be part of another document or may be a separate stand-alone document. It should define file-naming standards, projection and coordinate standards, data precision, layer or entity coding schemas and valid values, attribute table definitions, and annotation standards (font, size, and placement) when appropriate. The database design should also include definitions of the various fields to be used in the database, including acceptable rules for data capture.
- Metadata. Metadata tools are available to allow future theme based metadata to be stored in accordance with Federal Geographic Data Committee standards.
- Project schedule. An overall project schedule should be developed using tools such as Microsoft Project, Primavera, Excel, or a Web-based scheduling tool. This schedule should be updated and maintained throughout course of a project.
- Quality assurance/quality control procedures. A QA/QC document should be prepared. The document should outline vendor and client QC checks to be performed and acceptance criteria.
- Meeting minutes. Often overlooked, meeting minutes are key to ensuring all project related discussions are documented and jointly agreed to by all participants.

These documents will facilitate effective management of a conversion project. They will also be beneficial as technical and management staff change throughout the course of the project. As users get access to the converted data, they will also serve to enhance their overall understanding of the GIS data.

Many organizations elect to publish digital versions of each of the documents in PDF format to preclude more than one user editing each document, and to facilitate distribution of each document. One example of distributing the PDF document so that all project members have access to it can be found at http://www.eproject.com. At this site, project members, with one project manager to maintain and update the account through the Website, can have access to all aspects of the documentation process. Meeting minutes, database design updates and full releases, etc., can be posted on the site for all project members to review. Furthermore, once the document has been posted to the site, an automated e-mail is sent to all project members letting them know the second the document is online.

PILOT PROJECT. One of the key components of a successful utility conversion project is completion of a well-structured pilot project. The pilot

---

Figure 4.11 Example of a pilot project technique.

project provides an opportunity to finalize placement and capture rules, database design parameters, cartographic standards, and QC and data validation procedures. For most projects, a two- to four-sheet pilot project in an area that is representative of the project is recommended. An example of a pilot project technique is given in Figure 4.11.

Because the pilot project is critical to establishment of overall project procedures, it is important that an adequate amount of time is allocated to the pilot. Typically, three to six months would be required for a comprehensive pilot project.

Some organizations elect to complete a pilot project in-house, with production conversion work being completed by a conversion vendor. Completing a pilot in-house provides a training opportunity for staff and allows the specifications and design for full-scale conversion to be precisely defined.

Scrubbing procedures may also be developed during the pilot project. This may involve highlighting features to be converted, or clarifying anomalies and inconsistencies on sources. An example of a scrubbed map is given in Figure 4.12.

As part of the pilot project, production conversion work outside the pilot area should be minimized or preferably not started at all. Any design changes that occur as a result of the pilot should be clearly documented in the corresponding database design and conversion specification documents. It is also recommended that the pilot project be reviewed with the actual users of the data, in addition to being reviewed with those responsible for data maintenance and data creation. Getting this user buy-in is an important part of the pilot project and reduces the likelihood of major changes being required during the production process or, even worse, after production conversion occurs.

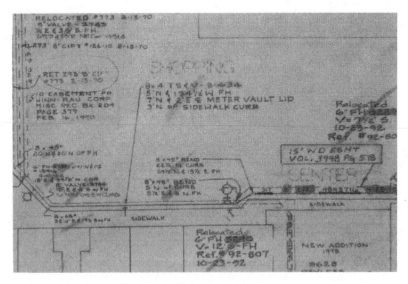

Figure 4.12 Example of a scrubbing technique.

PRODUCTION MANAGEMENT. Once the pilot project is completed, full production may begin. Some guidelines for effective management of the production process include the following:

- Receive data on an incremental basis. Incremental deliveries allow progress to be monitored and ensure that standards are followed. It is also recommended that the first production delivery area be relatively small in size so that any issues not addressed in the pilot will be able to be dealt with before production gets too far along.
- Data review and acceptance. In conjunction with the incremental deliveries, the review and acceptance process should follow an established schedule. Typically, a 30-day review period seems to work best.
- Work with vendor on QA/QC protocols. The QA process should be designed as a complementary process to the conversion vendor's production process. Tools developed by the vendor or by an organization should be shared freely. The objective should be to produce quality data and not to just catch errors.
- Error correction feedback. This is critical to project success. No matter how data is converted, there will be errors; documentation of errors and providing feedback in a positive manner allows the production process to

be altered so that the number of errors will be reduced with each successive delivery.

- Production facility meeting, Conduct at least one meeting at the production facility where the work is to be performed. This enables a user organization to get familiar with the overall process, and enables production staff to get a better understanding of a client's sources and requirements. Obviously, if the production work is being done by an offshore operation this may be difficult; however, even most offshore operations have a U.S.-based production facility where data is processed and reviewed before it is delivered to a client.
- Regular project meetings and conference calls. These meetings facilitate overall communication throughout the course of a project.
- Dedicate staff. One of the biggest obstacles to success is the lack of dedicated staff to manage the project. A utility should have a clearly defined project manager who is empowered to make decisions on behalf of the utility and who has access to the necessary technical and management resources to support the project.
- Plan for the unexpected. Most conversion projects encounter unexpected issues throughout the conversion process. To address these unexpected issues, contingency resources or plans should be included in a vendor contract to provide some flexibility in management.

PROJECT STATUS REPORTS. As part of ongoing project, management activities project status reports should also be provided on a regular basis. The frequency depends on the size and magnitude of the project. Typically, a monthly report is adequate. Ideally, information addressed in the project status report would include the following:

- Percentages of major activities completed,
- Summary of data production status,
- Description of current project issues and procedures,
- Activities to be completed over the next reporting period,
- Meetings held or planned,
- Listing of requested action items, and
- Listing of outstanding issues and action items.

Such reports must have a standardized format for easy understanding of project trends. A status-tracking map often accompanies this status report. With increasing regularity, companies are developing HTML links to project Websites that contain status-tracking maps and reports. In some cases, this information is updated on a daily or weekly basis to support status reporting.

The management and tracking process is key to overall project success. In a conversion project, it is very easy for a small increase in the time it takes to complete one step in the process to have a major effect on the overall project schedule and, in some cases, cost of the project. Effective project tracking serves to mitigate the risk of problems occurring in terms of schedule, costs, or quality.

Data Maintenance

Organizations have two choices regarding maintenance of the data throughout the conversion process. The first option is to maintain the facilities data during the conversion process. If a vendor is performing the conversion, this means that those copies of the new as-builts or scanned as-builts that show the work which has been completed, must be provided to the vendor.

The other approach is to complete initial conversion of all data based on documents that are "frozen" as of a certain date. After the conversion and all QA/QC and acceptance checks have been completed, the backlog of projects that were completed since the original data was converted is updated.

Many organizations elect to have all of their updates processed on-site using the same tools used during the conversion effort. This approach has the following advantages:

- It will serve to fully test the update and maintenance tools in an operational environment,
- Staff are able to get fully trained on the update process and tools,
- It is often easier for in-house staff to perform QA/QC on the data that has been converted, and
- It is often a more time-effective approach rather than sending the data back to production offices for conversion.

Some large utilities are also now outsourcing some of their utility maintenance operations to private companies. This approach is becoming more common for private investor-owned utilities (electric and gas). Outsourcing may involve periodically sending updated source documents to a maintenance contractor, or it may involve that maintenance contractors provide on-site staff augmentation services to assist with all or a portion of the updating work. It is also important to note that, irrespective of the approach, effectively maintaining data in a GIS environment requires a change in the work routines of staff, which often involves retraining and new responsibilities.

Data Migration

For many organizations, once their data is initially converted, in addition to maintaining their data, there is often a requirement to migrate data into different software format or database environment. Users migrating from Arc/INFO coverages to the Arc8 geodatabase, from computer-aided design (CAD) files to a GIS environment, and from DGN file to Oracle Spatial are all examples of migration activities. If this task is to be undertaken, many of the same steps involved in initial conversion will be revisited. The key requirement is to develop the data model that defines the relationship between existing features

---

and how they will be represented in the new data structure. Data migration should also involve completion of a pilot or testing phase, validation of the design against the application to be supported, and may involve extensive data cleanup or normalization before the migration effort. Currently, there are a wide variety of translation and migration tools available in the market. The key is to make sure the data being migrated is of sufficient quality and identify additional enhancements to be undertaken after migration.

CONCLUSIONS

The database development effort is a major undertaking for a utility. It defines how the data can be used and the level of integration with existing systems that can occur. It is often the most costly effort of an organization's implementation effort. The effort that goes into designing and developing a comprehensive database that is positionally accurate is significant. The benefits of a quality database greatly exceed the problems that can result from a poorly developed database.

Chapter 5
Enhancing Productivity through Applications Development

This chapter describes why geographic information system (GIS) applications are important and the way they are developed. An application is an applied use of a technology. For example, online bill or tax payment is an application of Internet technology, determining accurate locations of water infrastructure is an application of global positioning system (GPS) technology, and converting utility maps to hydraulic models is an application of GIS technology. No matter how noble a technology is, without applied use it is just a theoretical development. Applications bridge the gap between pure science and applied use. Applications make things easier to do, which saves both time and money. Continued development of new applications is critical to sustain the growth of a new technology. The GIS, being a new technology itself, cannot survive unless people use it to enhance productivity and develop cost-effective solutions. While GIS applications in the water environment are not new, getting beyond basic inventory and mapping functions is often challenging. This chapter shows how to put the GIS technology to productive use.

INTRODUCTION

Water and wastewater business is growing throughout the world. For example, the United States market for water quality systems and services had a total value of $103 billion in 2000. The two largest components of this business are the $31 billion public wastewater treatment market and the $29 billion water supply market. The water and wastewater engineering sector of this market experienced its third consecutive year of double-digit growth, increasing 14% in 2000 to $3.6 billion (Farkas and Berkowitz, 2001). More than 80% of all the information used by this business is *geographically referenced*, that is, a key element of the information used by water, wastewater, and stormwater utilities is its location relative to other geographic features and objects.

The typical local government office contains hundreds of maps displaying such information as municipal boundaries, property lines, roads, sewers, water lines, voting district boundaries, zoning areas, flood plains, school bus routes, land use, zoning, streams, watersheds, wetlands, topography, geology, and soil types to name a few. Paper maps, after all, have been the traditional method of storing and retrieving geographically referenced information. The shear number, range of types, and diversity of maps used by municipalities are evidence of the importance geographically referenced information plays in our day-to-day operations.

One of the biggest challenges of the rapid urban growth is managing information about maintenance of existing infrastructure and construction of new infrastructure. If information is the key to solve infrastructure problems, the first step of any infrastructure improvement project should be development of an information system. A GIS is one of the most effective tools for managing infrastructure information.

The real strength of a GIS is its ability to integrate information. It can organize the geographic information of a municipality or utility into one seamless environment. A GIS integrates many kinds of information and applications with a geographic component into one, manageable interface. A GIS offers integrated solutions in the areas of planning and engineering, operation and maintenance, and finance and administration. In the past 10 years, the number of GIS users has increased substantially. Exchange of data among GIS, computer-aided design (CAD), supervisory control and data acquisition, and hydrologic and hydraulic (H&H) models is becoming much simpler. Today GIS is being used in concert with applications such as maintenance management, capital planning, and customer service. Many of us are using GIS applications on the Internet and wireless devices even without knowing that we are using a GIS. These developments make GIS an excellent tool for managing wastewater and stormwater utility information and improving operation of these utilities. The time has come for all the professionals involved in the planning, design, construction, and operation of wastewater and stormwater systems to enter one of the most promising and exciting technologies of the decade in their profession—GIS applications.

Continued development of new applications is critical to sustain the growth of a new technology. Many water and wastewater utilities have digitized their paper maps in a GIS or CAD system. However, some of these utilities are now facing the "so what?" dilemma because automating maps and drafting efforts alone may not justify the financial support needed to maintain a GIS. To garner much-needed financial support, the GIS should support short- and long-term decisions required to run an organization. Geographical information system emphasis is now shifting from producing high-quality maps to enterprisewide critical applications. Thus, to realize the full potential of a GIS, it must be implemented on an organizationwide basis. The cost–benefit ratio of GIS increases with its functionality and applications. Geographical information system applications of automated mapping return a 1:1 cost–benefit ratio. The ratio increases as GIS use expands to more departments of the organization. Cost–benefit ratios of 4:1 can be attained when the entire organization shares GIS information and GIS applications are maximized (Alston and Donelan, 1993).

COMPELLING EXAMPLES OF GEOGRAPHIC INFORMATION SYSTEM APPLICATIONS

A recent survey conducted by the Geospatial Information and Technology Association indicates the following top 10 applications in the utility industry (Engelhardt, 2001a; GITA 2001):

(1) Landbase model,
(2) Work management,
(3) Facility model analysis and planning,
(4) Operations and maintenance,
(5) Document management,
(6) Customer information systems,
(7) Workforce automation,
(8) Regulatory reporting,
(9) Environmental testing, and
(10) Marketing.

Water, wastewater, and stormwater utilities were found to be focusing on work management, facility model analysis, facility planning, and pen/mobile computing. Some compelling examples of how the water environment utilities are using GIS applications are given below.

LOS ANGELES' AWARD-WINNING NATIONAL POLLUTANT DISCHARGE ELIMINATION SYSTEM APPLICATION. *Geospatial Solutions* holds an "Applications Contest" each year. The entries are judged based on interest, technology, and importance. The first place winner for the 2001 contest was a stormwater pollutant load modeling application for Los Angeles, California. As a condition of its National Pollutant Discharge Elimination System municipal stormwater permit, the Los Angeles County Department of Public Works is required to calculate and report pollutant load for more than 256 contaminants. The manual calculation of this information took engineers 250 working hours for an area one-quarter of the required study area. The county, therefore, turned to GIS to create an automated model for calculating pollutant load. Relying on an existing public works GIS, a team added several data layers to the system to integrate information about rainfall depth, drainage area, land use, imperviousness, and mean countywide concentrations for all 256 pollutants based on monitoring station data. With that information, the team developed a calibrated model that estimated pollutant loads with 95% accuracy. The project team used Environmental Systems Research Institute's (ESRI's) ArcView and Spatial Analyst for GIS analysis and an automated report generation process from another firm (Geospatial Solutions, 2001).

SAN DIEGO, CALIFORNIA SanGIS. The city of San Diego, California, was an early convert to GIS technology and is considered a leader in GIS implementation. With the motto, "We have San Diego covered" SanGIS (www. sangis.org) is a joint agency of the City and County of San Diego responsible for maintenance of and access to regional geographic databases for one of the nation's largest county jurisdictions covering more than 10 800 km^2 (4200 sq mi). SanGIS links 20 different city, county, state, and federal departments with a high-speed network of T1 (1.544 megabits per second) and T3 (44.736 megabits per second) lines. The SanGIS database covers 832 300 parcels, 35 700 streetlights, 501 schools, 64 eagle nests, and more. SanGIS contains 50 gigabytes (equivalent of 90 CD-ROMs) of data in 150 data layers. Their work crews use a sewer/water infrastructure management (SWIM) system to update maps in the field and create electronic work orders. The SWIM system uses portable GPS equipped pen-based computers to easily update maps and record information using light pens. The updated information is automatically identified and sent to map maintenance staff. SanGIS spent approximately $12 million during a 14-year period from 1984 to 1998 to collect GIS data. The conventional surveying approach would have cost them approximately $50 million (The San Diego Union-Tribune, 1998). The GIS–GPS approach saved the City and the County of San Diego millions of dollars.

THE MASSACHUSETTS WATER RESOURCES AUTHORITY GEOGRAPHIC INFORMATION SYSTEM PROGRAM. The Massachusetts Water Resources Authority (MWRA) provides water and wastewater services to 2.5 million people in 60 municipalities of the Greater Boston area. The MWRA is best known for its $3.5 billion dollar upgrade of the Boston Harbor wastewater treatment plant. The MWRA service area spans more than 800 sq mi that contains several treatment plants, 780 miles of large size pipelines, dozens of pumping stations, and tunnels. The MWRA recognized the potential for GIS to save ratepayers money and initiated a GIS program in 1989. The program grew from a tool to support general planning-level applications to an information management system that supports automated mapping, hydraulic modeling, site-specific analysis, maintenance, and facilities management. As the GIS data have been used repeatedly by individual communities to protect their water resources, the investment continues to pay off years later. For example, their geologic database "gBase" contains information on deep rock borings for the many tunnel and dam projects designed over the past 100 years. Ready access to this information guides current geologic exploration and enables MWRA to better locate new borings, which can cost $20,000 to $50,000 each (Estes-Smargiassi, 1998).

ADVANTAGES AND DISADVANTAGES OF GEOGRAPHIC INFORMATION SYSTEM APPLICATIONS

Today's GIS is limitless. Its applications are numerous. Their number is limited only by one's imagination and the availability of data. For instance municipal applications alone are tremendous, such as

- Water, wastewater, and stormwater operations;
- Permitting and code enforcement;
- Building inspection;
- Zoning;
- Parcel mapping;
- Tracking citizen complaints and requests;
- Grant applications;
- Comprehensive planning; and
- Routing.

Geographic information system applications can significantly reduce time and costs associated with conventional analysis and evaluation methodologies (U.S. EPA, 2000). As rewarding as GIS could be, it is not without some limitations and drawbacks. The advantages and disadvantages of GIS are discussed below.

ADVANTAGES. A GIS provides a spatial approach to organize information about customers and assets, such as sewer pipes, manholes, outfalls, pumps, and treatment plant equipment, of a sewer utility. Geographic information system applications help a wastewater or stormwater utility analyze spatial information about its customers and assets to improve planning, management, operation, and maintenance of its facilities. Organizations that have successfully implemented GIS have seen dramatic improvements in the way in which data is retrieved, analyzed, and maintained. These improvements are allowing municipal and utility personnel to collect information more efficiently, better perform routine activities, and make more informed decisions.

A GIS fosters better communication and cooperation among various stakeholders (e.g., community leaders and the public) of a wastewater or stormwater construction or improvement project. Many people learn better with maps than they do with words or numbers. A GIS can be used to communicate with different audiences using visually different views of the same data. For instance, three-dimensional (3-D) plan views of a sewer system improvement project can be used for presentations at town meetings to graphically illustrate necessary improvements. Because GIS is a visual language, it is an excellent communication tool for visual learners. Figure 5.1 shows a 3-D layout of a sewer system created with GIS.

Geographic information system applications increase efficiency, save time, and translate into saving money. A GIS allows the performance of routine work, such as keeping records of maintenance work or customer complaints, to be more efficient. Geographic information system tools are becoming user-friendly and easier to use. Local governments, utilities, and their consultants are using GIS to analyze and plot geographic solutions in a fraction of the time previously required. Geographic information system functionality is being imbedded in other applications. The benefits of GIS include increased

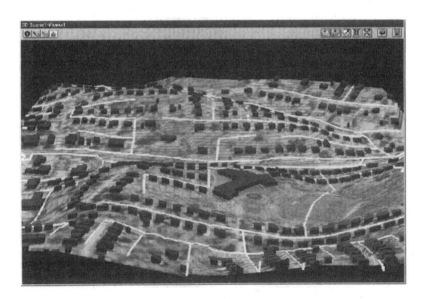

Figure 5.1   3-D view of a sewer system created using GIS.

productivity, quicker turn-around, and enhanced integration of applications because of integration and centralization of mapping and attribute databases that are current.

By using geography as the common denominator, GIS permits data from a wide range of disparate sources to be combined and analyzed. By using GIS software, the resulting information can be kept current and accessible for use in an enormous number of applications. Therefore, the foremost benefit of GIS applications is their inherent ability to integrate and analyze all spatial data to support a decision-making process. This integration power makes the scope of GIS applications almost infinite. A GIS provides the uniformity of data usage and flexibility to test and evaluate multiple scenarios. Use of a common database eliminates differences in presentation, evaluation, and decision-making based on using different forms and types of data. A GIS provides the opportunity to conduct sensitivity analyses appropriate for the level of accuracy of the input data. This allows engineers, planners, elected officials, and the public to focus on the effects and analysis of alternatives rather than accuracy of data. Geographic information system applications can significantly reduce time and costs associated with conventional analysis and evaluation methodologies. After the planning and decision-making phase has been completed, the GIS can continue to support the implementation phase of a project by tracking the success and failures of alternative approaches. Plan performance tracking and testing of new approaches is based on new parameters, new information, and new conditions within or outside the study area (U.S. EPA, 2000).

DISADVANTAGES. As with any new technology, GIS has its drawbacks (U.S. EPA, 2000). The first issue is the substantial time and cost required to compile and analyze the necessary data. A survey conducted in 1988 by a GIS consulting firm on GIS management revealed that cost is the largest obstacle to overcome in convincing management to fund GIS projects (Alston and Donelan, 1993). Unfortunately, because of the difficulty in quantifying qualitative (intangible) benefits such as more informed decision-making and enterprise-wide data access, many GIS investments may not be amenable to standard cost–benefit analysis.

Though the advantages of GIS applications are dramatic, the failure to effectively implement GIS can lead to disappointment and disillusionment with the technology. Improperly designed and planned GIS applications will result in costly and time-consuming efforts. High initial costs will be incurred in purchasing the necessary hardware and software and for ongoing maintenance. Geographic information system applications are disadvantageous when one fails to define a vision, understand the vision requirements, define the tools needed to attain the vision, and select appropriate technology to integrate those tools. With powerful computers and user-friendly mapping software, unintentional cartographic self-deception can be inevitable (Monmonier, 1996). Consequently, while GIS should be approached enthusiastically, implementation must be carefully planned and effectively accomplished. Some of the factors that must be considered and addressed from the outset include immediate and long-term applications, data accuracy requirements, budget, staffing, and staff training requirements.

The GIS learning curve; privacy issues; shortage of skilled personnel; and cost of developing, maintaining, and operating GIS databases are some of the drawbacks of a GIS. However, as *The San Diego Union-Tribune* (1998) reports, "Those who overcome such hurdles soon find GIS applications breathtaking in scope and far reaching in the potential to affect, if not shape and change, everyday lives."

*F*UTURE GEOGRAPHIC INFORMATION SYSTEMS APPLICATIONS AND TRENDS

In the mid-1990s, GIS technologies were relatively new and still near the lower end of the growth curve. Geographic information systems are experiencing rapid growth as information management tools for local and regional governments and utilities because of their powerful productivity and communication capabilities. Though GIS has come far, many changes are still to come (Lanfear, 2000).

Most GIS experts believe that the future of GIS looks bright thanks to recent advances in the GIS-related technologies. The past decade was marked by primitive data models and highly skilled end users. The future will bring advanced data models, user-friendly GIS interfaces, and casual to moderately skilled end users. As GIS technology moves into mainstream society, more clients will demand new GIS applications. Geographic information systems have been traditionally used in planning stages of a project. New trends will extend the GIS applications into the construction and project management phases.

Based on the predictions of some GIS industry experts (Engelhardt, 2001b; Geospatial, 2001; Lanfear, 2000) the following trends are expected in future GIS technologies and applications:

- Spatial data will become more widely accepted, unit costs of spatial data will fall, and GIS applications will become more affordable.
- Location-based services combine GIS applications with user-friendly mobile devices to provide needed information at any time or place. It is estimated that, by 2003, 50% of U.S. workers will use mobile devices, wireless Internet use will surpass wired use, one billion people will use a Web-enabled mobile device, and 80 million of those devices will be equipped with location capabilities, either through GPS, a wireless network, or a hybrid solution (Barnes, 2001). Location-based services devices paired with wireless technologies will provide "on-demand" geospatial information.

- The explosion of spatially enable consumer products will benefit the entire GIS applications industry. Soon, GPS will reside inside cellular phones. By 2005, 62% of the world population is expected to have wireless Internet connections. A new generation of low-cost consumer GPS products and spatially enabled mobile phones and personal digital assistants (PDAs) will provide consistent accuracies of 10 m or better—sufficient to navigate people to their destinations. Location information via GPS will become as common as timekeeping on a wristwatch.
- Sophisticated data-processing techniques will be developed to generate user-friendly and customized information for decision making. Advance datasets will allow us to create more comprehensive computer models. Anticipatory modules will be developed in software and databases to evaluate alternative future scenarios, review implications, and choose a preferred future course of action. Mappers will be able to morph a current image into various 3-D images to show the effects of alternative choices. Maps of the future will let us stand in the future and look around.
- Light imaging detection and ranging will become a true GIS tool and produce new varieties of GIS data products.
- Commercial remote sensing companies will be selling more than satellite images—they will be offering subscription-based monitoring services for change detection and vegetation indices, and so on.
- As the bandwidth of the Internet improves, it will enable us to deliver information-rich geospatial content directly to end users.
- Geographic information system databases will be connected with real-time sensor data inputs, which will provide new opportunities for mapmakers and resellers of GIS products and services. New markets will be developed to expand the selection of datasets.
- Evolution of a spatial network query language from the structured query language (SQL) for serving GIS data over the Internet, wireless systems, and PDAs.

This large list of potential applications clearly indicates that future GIS applications will be even more exciting than those that have been developed in the last decade. The effect of these new applications will be profound. They will be inexpensive, user-friendly, and ubiquitous. They will support integration of GIS and related technologies in ways unlike anything current wastewater and stormwater professionals have ever envisioned.

APPLICATION DEVELOPMENT PROCESS

Although it is possible to create a GIS application starting from scratch, GIS applications are most often developed by extending the core capabilities of a commercially available GIS software package. Creative application developers can find something useful in a GIS package and turn it into something innovative. The two methods for developing GIS applications are GIS customization and programming. Geographic information system customization generally changes the default GIS user interface by adding new tools and menus that perform news tasks. The customization capability of a GIS software package is the key to developing applications using this method. Customization is appropriate for small applications. For larger applications, new computer programs must be written and linked to a GIS database. Many applications use a combination of both methods.

Basic GIS applications and customization, such as adding a new button or a menu, may be achieved without programming. However, advanced applications, such as creating a link to a computer model, almost always require some programming using a scripting language. A scripting language is a programming language that is (typically) embedded in another product, such as Microsoft's Visual Basic for Applications (VBA) or Autodesk's AutoLISP. Scripts or modules are small computer programs written in a scripting language. It is important to note that end users do not need to know how to program—all they have to do is "point and click".

Finally, it should be noted that in-house development of applications is not necessary. A large variety of third-party applications from facilities management to H&H modeling are available for efficient operation and management of water and wastewater systems. A utility should determine whether its wants to find an off-the-shelf, predesigned application or whether it wants to build a custom application in house. Water utilities that do not have the resources or technical expertise necessary for developing in-house applications might find it easier to purchase commercial, off-the-shelf applications.

Environmental Systems Research Institute (ESRI) is one of the world's leading GIS software companies. As an example, the following paragraphs discuss how to develop GIS applications using ESRI products. Products by ESRI are used here for illustration purpose only and do not imply an endorsement. Other GIS packages that allow interface customization and have a scripting language can be similarly used for developing GIS applications.

The latest versions of ArcInfo and ArcView (versions 8 and higher) can be customized using VBA. Older versions of ArcInfo, PC ArcInfo, and ArcView are customized using Arc Macro Language (AML), Simple Macro Language, and Avenue scripts, respectively. In ArcView 3.x, scripts must be compiled and linked to GIS software before they can be executed, which can be cumbersome for casual users. Alternatively, a set of scripts can be converted to an "extension" for faster and more user-friendly installation and execution.

Avenue was the native object-oriented scripting language for ArcView 2.x and 3.x built (integrated) into ArcView. To the extent that these earlier versions of ArcView continue to be used, application developers and programmers can use Avenue to modify the user interface, build custom tools, and develop solutions for specific applications. Avenue allows simple customization, such as the addition of a new tool to complete a new task or the creation of complete turnkey applications. Avenue's full integration with ArcView 3.x benefits the user in two ways: first, by eliminating the need to learn a new interface and,

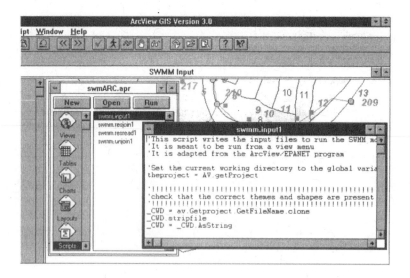

Figure 5.2   ArcView GIS application development environment.

second, by letting the user work with Avenue without exiting ArcView. Figure 5.2 shows the integrated Avenue programming interface of ArcView 3.x. Avenue developers can also customize the ArcView 3.x interface using ESRI's "Dialog Designer" extension. Like VBA, Dialog Designer allows programmers to call up Avenue scripts by clicking on buttons.

A sample Avenue script for displaying digital video files in ArcView is shown in Figure 5.3. This compact script uses ArcView's "HotLink" function and Windows Media Player software. It can be used to launch a video file by clicking on a feature that has been linked to the video file. For example, as shown in Figure 5.3 this script can be used to display Audio Video Interleaved (AVI) or Moving Picture Experts Group (MPEG) files of sewer system television inspection videos simply by clicking on a sewer pipe in ArcView.

Applications for ArcInfo 7.x and earlier versions are developed using its scripting language, AML. Designed for these versions of ArcInfo, ArcTools is an AML-based graphical user interface composed of sets of tools that represent individual ArcInfo commands or sets of commands. It provides generic

```
MovieHotLinktoMediaPlayer
theVal = SELF
if (not (theVal.IsNull)) then
  if (File.Exists(theVal.AsFileName)) then
    System.Execute("C:\Program Files\Windows Media Player\MPLAYER2.EXE "+theVal)
  else
    MsgBox.Warning("File "+theVal+" not found.","Hot Link")
  end
end
```

Figure 5.3   Sample avenue script.

messaging and file-browsing tools useful for general application development. Initially released with ArcInfo version 6.1.1, ArcTools allows ArcInfo customization and makes ArcInfo easier to use.

For software developers who want to include mapping capabilities for their applications, MapObjects, ArcObjects, and ArcIMS are appropriate. MapObjects Version 2 is an ActiveX control, with more than 45 programmable ActiveX Automation objects that can be plugged into many standard Windows development environments such as VBA, Visual C++, Delphi, and PowerBuilder. MapObjects supports a wide variety of data formats (e.g., shapefiles, ArcInfo coverages, and computer-aided drafting and design [CADD] formats). ArcObjects is Microsoft's Common Object Model (a standard for independently developed components to communicate with the operating system, often referred to by its acronym, COM) based framework that allows developers to enhance the ArcGIS Desktop user interface and extend ArcGIS data models.

APPLICATION DESIGN

Software engineering is the term used to describe the discipline of creating computer applications. However, the process of building a computer application can be very different than the engineering required to build a structure.

Developing applications for GIS systems used to exclusively center around the GIS package that would serve as the platform for the application. Fortunately, the convergence of many technologies has provided a more common platform on which to build applications. This will lower the cost of development by allowing for less specialized expertise, requiring fewer specialized tools, and promoting the reuse of application components. The details of the application design process can be applied to any kind of application (GIS or otherwise). Because of the general nature of the subject, many texts cover this area in detail (for example, see Kruchten, 1999; Royce, 1998). Specific issues related to GIS applications will be noted when necessary.

THE APPLICATION LIFE-CYCLE. Project development tasks are coupled to the object or goal of the project. For example, a project developed to create software will differ fundamentally from a project to create a bridge or develop a marketing plan for a product. Many water and wastewater professionals have experience in building long-lasting physical structures. A software product, typically having such a short life span, is commonly built using a development cycle approach, acknowledging the rapid changes in underlying technology or associated data that often drive the need for changes to the product. In this sense, the project approach is not linear but can be thought of as a spiral (Boehm, 1988) as shown in Figure 5.4, moving to a higher level of function through each rotation, mimicking a "version/release" model. In this model, common in commercial software, a product is described with specific features and supported when released to the users and assigned a specific identifying

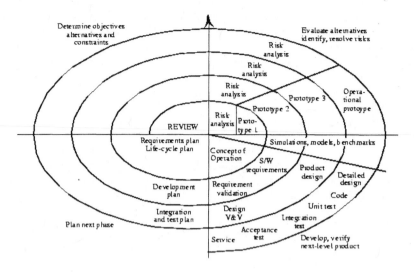

Figure 5.4 Boehm's spiral model for the application life cycle.

version (typically a numeric indicator, for example, version 1.2). Any subsequent changes to the application (whether to fix problems, add additional features, or both) are released with higher version numbers. Many commercial vendors will continue support for older versions for a limited period of time.

The waterfall method of software development is a linear method characterized by extensive documents at critical milestones. This method is executed in sequential fashion, making it impossible to incorporate feedback from downstream tasks. A method that includes prototyping, which allows for iterative development and incremental progress, is recommended in developing applications.

The following sections describe various steps used in the development life cycle. These steps focus on the technical aspects of the cycle, not the managerial tasks. Depending on the organization, various milestones will require approval of expenditures, cost–benefit analysis, or other oversight and review activities.

Needs Requirements. As is typical of all projects, the needs of the product are determined and written into one or more specification documents. To fully understand the needs, the project team should be aware of existing data, process flow, and infrastructure. It is within the existing context that users often articulate what is needed with a new (or upgraded) system. The requirements at this point should be in the form of what are commonly referred to as "business requirements". These statements answer the question of "what is to be done" rather than specific implementation ("how it is to be done").

Business requirements should be readable so that any stakeholder can understand the purpose for the project. These can be in the form of problem statements or other such language that can be understood by stakeholders or management-level decision makers. In most cases, the objective should be to reduce operating costs, improve service, or reduce risk. The requirements should not involve designing the solution (for example, "We need to reduce our customer's average delivery time to one week", not "We need to migrate our order-entry system from the mainframe to a client-server system"). These high-level requirements can serve throughout the development cycle as a reminder of common objectives. When focusing on detailed tasks, a periodic review of the business requirements may help from drifting away from project goals.

Specification. An initial specification is written at a detailed level. These functional requirements will provide the basis for designing the solution. These can be written after developing "use cases" using Unified Modeling Language (UML) (see the Unified Modeling Language section later in this chapter). Specific functions required by the system should be written in such a way that a test document or acceptance document could be produced from the same list. Therefore, each requirement must be "testable" when the system is complete. Words such as "good" or "fast" should not be used as they are more subjective descriptions and can lead to disagreements over judgments.

The functional requirements document is not intended to communicate to management or higher-level stakeholders. Instead, it is a tool to be used between the system developer and the system's ultimate user. If the development is for a specific client, one or more key members of the client's team should be assigned to help develop these requirements.

Applications will have dozens (perhaps hundreds) of requirements. For example, a sewer inspection tracking application may use a map interface to track the status of cleaning crews. The inspection planner may have a requirement such as the following:

When at least one pipe segment has been selected, the user will be able to change the status of all selected pipe segments to a status code. The list of valid status codes is

<null>—not planned for inspection
P—planned for inspection
F—planned for inspection, needs flushing
I—in progress (have started inspection but remains to be completed)
A—inspection halted/abandoned
C—inspection complete

Note that this example does not deal with how pipe segments can be selected—this would be stated in a separate requirement. If the user felt that changing the status was an event that required confirmation, he or she may wish to rewrite the specification as follows:

When at least one pipe segment has been selected, the user will be able to change the status of all selected pipe segments to a status code. After the new

status code has been selected, the user shall be given a confirmation state-
ment listing the number of segments for which the status code will change.
The user shall be able to cancel from this point without any changes in status
code or shall be able to commit the changes to the pipe segment database.

Solution Domain. When the project team is assembling specifications for the
system, the type of solution will typically be discussed. There may be many
alternatives for addressing the solution. For example, a transportation corridor
crossing a river may require a ferry, a bridge, or a tunnel. Similarly, there are
alternatives to assess in developing computer applications.

A significant part of application development rests in investigation and
selection of an architecture or framework with which the application will be
built. Attributes of the architecture include

- Target operating system,
- Development environment,
- Customization of other available software, and
- Database platform.

Based on the application requirements, there may be many choices and many
decisions regarding the architecture. Conversely, an application for which
requirements closely match an existing product will involve fewer decisions.

In the area of GIS, the selection of the database platform will have the
largest effect on the selection of the development environment and the components available. In many cases, this selection has already been made as part
of an organization's strategic information management plan. However, there
exists a wide array of solutions in managing geographic information. What
has traditionally been a file-based domain, where proprietary software is used
to perform all the functions of a database, is moving to the use of more standardized (i.e., SQL-based) relational database products. There are many advantages to delegating various data management responsibilities to mature database
products, including security management, backup and recovery, and support
for multiple users. From a developer's point of view, the database is treated as
a service provider rather than a folder containing a group of files. However,
the implementation of 3-D geographic data types into a standard relational
database, with traditional tables designed for storing basic text and numeric
data, will have a significant effect on the development environment in building
applications. To further challenge the developer, many of these products are
changing rapidly, and industry consolidation can render even the most elegant
solution obsolete.

The future viability of many of these solutions rests on the use of standards.
A development team or system architect that is proposing a solution based on
these systems should research this aspect closely.

A development environment can be simply the scripting language built in
to an existing end-user product. However, it may be more complex. Consider
a project that uses a Web-based interface that requires development of specific
data access modules. Many third party products have emerged to serve mapping

applications over the Web. Which of these is appropriate to deploy? If additional components are to be built using COM, there again are many development environments available in which to build components. Be aware of tradeoffs between environment features, run-time performance, up-front licensing costs, upgrade costs, and platform availability. A number of questions need to be considered before committing to a specific product

- Is the available staff familiar with a particular platform?
- If not, what training needs will be required to gain the benefits of the environment?
- If staff turnover is a problem, is the environment sufficiently known in the pool of possible replacements, or will you need to send each new programmer to training?
- What is the licensing model for the environment, and what are the ongoing costs to maintain and upgrade the programming environment into the future?
- Will the chosen platform create programs that require extensive installation or maintenance issues on client hardware or on the server?

Through the research conducted before development begins, a long-term, cost–benefit comparison can be made.

A common question that arises early in the phases of software development is whether to purchase (and optionally extend) another application or build the required functions using development tools. This question does not necessarily apply only to the primary purpose of the application; it can also apply to specific functions that may be purchased as components that can be programmed as part of the solution. These decisions vary across a spectrum of buying products and building modules. For example, charting functions can be programmed into a Microsoft Windows application using direct references to the operating system run-time libraries to manipulate objects on the user's screen. However, unless the project team has access to existing source code that is sufficiently generic to address this application, many days or weeks will be spent developing this capability. Many commercial development tools are available that can provide the same functions. These components have predefined behaviors that may not be exactly what is required, but often the benefits of faster deployment and less maintenance outweigh the elimination of a specific (often esoteric) user requirement. These kinds of choices may exist throughout the application development process.

For many GIS applications, the user's environment dictates the database, the development environment, or both. There may not be too many decisions to make on architecture, but as more service-based applications become available, there will be more choices for the development team to make. These "buy versus build" decisions lie on a spectrum, with one end ("buy everything") essentially establishing the project as one of system integration.

When selecting commercial products for development environments (or middleware or specific components), be sure to evaluate which elements are customizable and which elements are fixed. For example, can the toolbar on the Web-mapping applet (a little application or program) be customized? There also may be run-time licensing costs for components. Another aspect to consider is the maintenance and upgrade costs for third-party component software. If the application platform is changed at a later date, these components may or may not be amenable to this change, and this complexity weighs into the cost–benefit decision.

Prototype Development. As proof of a concept or simply to elicit more detailed requirements from users, a prototype is often developed. The prototype, by definition, will not be used in production. However, it may be the basis for production modules. It can be built using a development environment different than that selected to build the final product. The prototype may be used to learn key limitations of the development environment or test concepts discussed during specifications.

Prototype Review. If the development is for a specific client, the prototype is shown to members of the client's team to help refine specifications. This may be an iterative process over several weeks.

Final Specification. The final specification, written using UML or other language that can be used as a test document, is prepared and subsequently agreed upon by the various stakeholders.

Final Development. Before entering final development, it is recommended that the requirements be dated and published for the project team. Typically, additions to the requirements will emerge from the development process, even after the specifications are deemed "final". The management of these changes can have a significant effect on the successful completion of the development project. In some organizations, an entire discipline ("change management") has been developed to deal with changing requirements. The acknowledgment that additional requirements can be addressed in later releases of the program (under separate budgets) is one aspect of managing the software development process.

Through a combination of system architecture, data modeling, and selection of available components, the application's required modules are defined and assigned to project teams. This combination may be an extensive task if the project is a stand-alone application or it may be trivial if the project is simply the extension of an existing product. Regardless of project size, the process of development first involves the design of discrete modules that will serve a specific function for the final product.

Good module design is based on completing one and only one task in a module and on eliminating any coupling between tasks and components. There are many books and articles describing distinct "layers" related to the decomposition of the application. These layers typically consist of

- Data layer—this is where the communication takes place between data and the application. Functions for connecting to the database, reading data, creating new records, deleting, or modifying records would exist in this layer.

- Business rules layer (analysis layer)—this layer embodies the functions listed in the specifications that relate to the workflow being performed.
- Presentation layer—this is where the interaction with the user takes place.

Note that in projects where an existing application is being customized, the design task of assigning modules to these layers may not apply.

One common example to reduce coupling is to establish database locations in a single module. In an older architecture, this may involve network paths to data files. By consolidating these assignments, a migration to a new network server can be accommodated by modifying this single module.

Another example of coupling may be one in which a specification calls for the user to enter an inspection record for a facility. The specification calls for the user to also update the preventive maintenance record for this same facility (because it occurs at the same time as the inspection). However, in the module assignment for this specification, the update to the inspection records should be coded separately from the module to update the preventive maintenance record. In this manner, future changes to business processes can be cleanly handled.

By decomposing the problem into manageable pieces, an assignment list and schedule can be developed and revised as progress proceeds. Regularly scheduled feedback to users and clients during the course of development is essential to ensure confidence that the project is on course.

As each module is enumerated in the project task list, conduct research to see if a production-level solution already exists. Depending on the circumstances, there may be modules available that were used in a previous project. A third-party commercial package might be available. A good design will strive for generality (to promote reuse) and maintainability. Only after determining that no feasible alternatives exist should coding be undertaken.

Testing. Each module is tested against the input and outputs. At integration testing, the combined modules are tested against the specifications. The various aspects of this stage are discussed later in this chapter.

Deployment. In the iterative and incremental development methods, a deployment stage will emerge gradually. However, be sensitive to end users' desire for a distinct signal for product acceptance. The project plan should allow for a stage of testing that is considered "final" for the specified version of the product. When the system meets all of the specifications listed for this version, any further problems that need to be resolved should be termed "maintenance".

Maintenance. Typically, software has a short life span. Before implementation, a plan for support of the application should be in place. Depending on the user base (internal or external), a small team, large team, or single individual may be placed in charge of fielding questions and "bug" (computer program coding error) reports from users. Inevitably, some bugs or questions are really ideas for improving or upgrading the application. The cycle of "version-release" should be used to incrementally update the program. Even if an application is "perfect", the underlying operating system, development environment, or database software will become outdated. At the very outside, the hardware will fail, requiring replacement. Eventually, manufacturers of the hardware or peripherals will cease to support the underlying operating system. It is, therefore, imperative to develop and follow a maintenance plan for the software or expect people to stop using it. Such a maintenance plan should anticipate items such as

- New output devices (e.g., plotters) that should be used,
- New versions of software used to create the application (upgrades),
- New versions of the underlying operating system, and
- Addition of new users.

OBJECT-ORIENTED DESIGN CONCEPTS. There are many resources available on object-oriented development techniques (for example, Booch, 1994; Coad and Yourdon, 1991). While it has become widely accepted that using these techniques has strong benefits, there are a variety of methodologies and frameworks that are considered "object oriented". Furthermore, the extent to which some of these methods can be adhered to will depend on the development environment (i.e., programming language) used to develop the application. However, just because a language does not support all of the features of an object-oriented environment does not mean that certain object-oriented techniques cannot be implemented. A FORTRAN (computer programming language) module can be written to create and destroy objects and can contain procedures (functions) to implement methods and manipulate properties of those objects. Similarly, a language that is considered object oriented (such as C++) can be hijacked to "break the rules" of object-oriented methodology. In summary, object-oriented techniques often are implemented more in the minds and disciplines of the developers than in development environments.

Object-oriented methodologies have been developed to help manage the complexity of software development. Each object represents a combination of data and methods. The only avenue to changing the state of the object is through its published interfaces. A "class" is the template for an object, which defines its data structure and functions (or methods) available for the object. One advantage that this provides is that interdependence, or coupling, of objects is reduced. This allows the development process to stage upgrades of the objects independently as well as allow for the assignment of object programming to multiple teams.

Using object-oriented methodology means that the application development process will define the classes and the interaction among the various classes (a "class diagram"). In this manner, developers (and clients on the project team) can discuss relationships among various objects using real-world terms. This connection between the software system and the problem domain is one of the greatest benefits of using object-oriented methods.

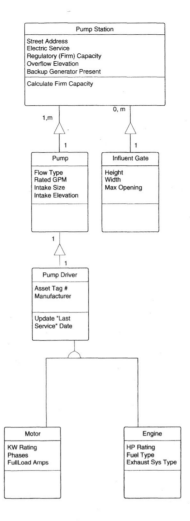

Figure 5.5   Class diagram example.

An example of a simple class diagram is shown in Figure 5.5. The notation being used in this figure is Coad–Yourdan, one of several notations available to describe class diagrams and object relationships. In this notation, each class is shown as a rounded box with three vertically oriented text areas. The top area simply states the name of the class. The middle area is a list of attributes (or properties) that are associated with each instance of the class. For example, the pumping station class has an attribute of "Street Address". Of course, each separate instance of pumping station will have a different value for this property. The bottom third of the class diagram box lists the methods (or services) that are available for each instance of the class.

Relationships among classes are either "whole–part" or "general–specific". The former designation is used when some classes are subsets (or pieces) of a larger object. In this example, the pumping station is made up of one or more pump objects. The triangle symbol is used on a line connecting the two objects, with the triangle shape pointing to the "whole" object (in this case, the pumping station). Numbers next to each end of the connecting line indicate whether the relationship is one-to-many, one-to-one, or many-to-many (also called "cardinality"). In this example, the pumping station can have one pump or many pumps ("1,m") and a pump can be associated with one and only one pumping station. Contrast this with the influent gate, where a pumping station can have zero to many ("0,m") influent gates.

A relationship that is "general–specific" is used to describe different types of objects that may share some aspects but not other aspects. In this example, a "pump driver" refers to the object that is moving the pump impeller, either an engine or an electric motor. Regardless of the type of driver, it will contain certain attributes such as the name of the manufacturer, but other attributes (such as the fuel type of the engine) will not be applicable if it is an electric motor. A semicircle is used when branching between specific types of an object.

UNIFIED MODELING LANGUAGE. One of the challenges in building large applications is in properly communicating throughout the project team. To establish a reproducible method for building systems, a language has been developed by a consortium operating as the Object Management Group. Although this UML covers a wide array of development tasks, among the most important is the development of use cases, which are used in the development of user requirements.

As noted earlier in the chapter, the functional specification of a system is used to communicate between the developers and end users. The UML use cases provide a method for documenting interactions between external systems (called "actors") and the system. Initially, interactions can be described in a coarse sense, with details added incrementally, until the user is satisfied that all interactions have been described. It is important that each use case focus on the business issue being addressed (although this should be as specific as possible). It may consist of a single transaction between the user and the system or it may be a defined sequence of transactions.

In many circumstances, the definition of the use cases can help establish the boundary of what the system will do. The user of the system should be able to clearly see from the description (or diagram) the goal of each transaction. Disputes will still occur as to how the goal is reached, but these can be addressed by further detailing the transaction. For example, if the goal is to add a street label, this can be stated very simply in describing the interaction between the user and the system "user selects a single street → user adds a street label".

In all but the most trivial systems, there will be relationships between multiple use cases. In UML, there is specific language that describes these relationships. "Extends" is used to describe the relationship between a more

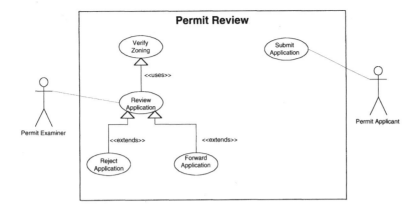

Figure 5.6   Use case example.

general use case and a specific use case. "Uses" is used to describe a common function, embodied as a use case, which is shared by other use cases.

In Figure 5.6, an example of a use case diagram is shown on a conceptual level. This is an appropriate place to start when describing a "system". Note that a rectangular boundary is used to define what is part of the system and what is not part of the system. In this example, the application submission by the applicant is shown as part of the system.

There are two actors in this example: the permit applicant and the permit examiner. There are five use cases shown (note the oval shape used to denote a use case). The "Review Application" use case has references to three other use cases

- A "uses" relationship with "Verify Zoning". Part of the review process involves checking the current zoning against the permit application.
- An "extends" relationship with "Reject Application" and "Forward Application". These use cases occur as a result of the review process.

Each of these cases can then be described in greater detail as the user and designer discuss the details of each process.

ADDITIONAL REQUIREMENTS. As mentioned above, the design process depends on the documentation of functional requirements. To properly document all of the requirements, it is recommended that these topics be discussed with end users and ensure a complete list of requirements

- Performance requirements,
- Diversity of client platforms—"Preferred" software platform,
- Data concurrency requirements, and
- Network infrastructure.

Each of these elements is described in more detail below.

Performance Requirements. The general subject of requirements has been discussed above in The Application Life-Cycle section. However, performance requirements are discussed here only because this topic is often left out of the requirements process. This does not mean that performance requirements are absent from users' minds. On the contrary, each user has a concept of what is acceptable for response from the application. Dealing with these variations is important, and important to confront early in the project. The selection of application architecture can often have a significant effect on application performance.

The selection of a development environment cannot be made without assessing effects on application performance. Does the application require many user inputs before solving a complex problem (in which case, the user may tolerate a long wait for the solution to appear) or will the user expect an interactive response (such as a pop-up appearing when the mouse cursor scrolls over an object)?

There often exists a tradeoff between long-term application viability (in the form of portability and maintainability) and performance. Architecture and development platform decisions made early in the design process should be tested on typically sized databases to ensure that, even if performance requirements are not explicitly written down by users, application performance does not become problematic as deployment nears.

Diversity of Client Platforms. A brief review of potential client environments should be undertaken at this step. Is the application being deployed at a single office or will it be used in several locations? In the case of a single client/single office deployment, is there a "preferred platform" already established?

In the commercial market, a decision must be made as to minimum client hardware resources, including central processing unit, memory, and disk space. If specific peripherals are required, this should be determined at the outset. For example, will color-printing capability be assumed or will black and white output be designed into the solution?

Data Concurrency. Although many useful applications exist that simply retrieve relatively static datasets (perhaps updated monthly), it is now increasingly common to see applications built that retrieve and edit data that is constantly changing (such as property transactions) or in real time (such as GPS tracking data).

Perhaps of special note is the transaction nature of databases. For example, you may want to have a database of pumping stations and, within this database, is a list of how many pumps are in each particular station. When a station is upgraded and more pumps are added, the field used to store the number of pumps is updated. If one were to ask, "How many pumps were in that station during the 1987 hurricane?", the design may not be able to answer this question. In some cases, the benefit of historical tracking does not justify the added expense of developing this capability. This is different from a property database, in which the ability to trace ownership and property short-plats (subdivisions) is often required.

Network Infrastructure. Another aspect of application architecture is the location of the database and what network is required to support data requests from clients. The application design should account for required interfaces and state clearly what assumptions are being made in this area. An increased emphasis is being placed on network security, and the use of routers and firewalls is common. The physical location of the user may have an effect on the ability of the application to succeed.

COMPONENTS OF APPLICATION DEVELOPMENT

A custom application differs from general software in the way its activities are specified, whereas general software provides tools that support any query on any data. A custom application may provide a specific tool that executes exactly one query on exactly one database. The value of the custom application is that it sacrifices the generality of the base software in favor of specific business processes that are important to the application user. Because of this, a successful application must be developed in a manner consistent with three critical factors:

- The technical environment must be defined accurately enough that what the application does is what it is supposed to do—no need to develop a tool that sorts by first name or last name if there is only a single name field.
- The capabilities of the custom application must be consistent with the procedures to be used. If new service workorders are to be generated by the GIS but new service points cannot be created without an account and an account cannot be created without a workorder, then this organization is not going to get a lot of new business.
- The custom application must be accepted by many people. If maintenance crews cannot use it, system people cannot reinstall it, or the department head cannot afford training time required, then the application will be an operation failure, even though it is a technical success.

Finally, the development process must have flexibility. Few organizations have specified and developed a custom application quickly and precisely enough that database, operational, or personnel changes were not necessary after development has begun. In fact, it is most accurate to consider custom application development as being an almost continuous process; this is not to say that a constant level of resources is required throughout the project, only that the development plan must be flexible. One important technique to bridge between the lesser flexibility of resource availability and the greater variability of resource demand is to segment the application development project into discrete pieces. Each piece can involve fewer people, less time, and a more specific set of dependencies than the project as a whole. Furthermore, a successful component piece, with whatever its inherent value is, can empower progress in related areas as the staff involved gain confidence and users support the project.

TECHNICAL ENVIRONMENT. Without having input information that is complete, correct, and structured, an application will generate results that are incomplete, incorrect, and inappropriate, but not necessarily noted as such. Data are paramount. A complete ensemble of software and hardware must perform in concert with the application and all of the people who interact with it.

Data. All considerations that apply to data systems typically are just as important in a GIS; unique identifiers, related records, maintenance responsibilities, and backup–restore processes are no less significant (or demanding) than in a traditional information system. However, the spatial aspect of GIS data adds considerations that have no analog in standard relational database. Two important examples of this are

- Unit definition: In a traditional database, we may model a pipe as a record with material, size, installation date, project number, location description, etc. In a spatial model, on the other hand, we may require that the spatial representation of our infrastructure support tracing along from one pipe to the next, including service taps and lines. This idea is illustrated in Figure 5.7. This model represents two pipes as four features: 33506, 34164, 33505, and 33507. This raises a question: Does each segment between taps constitute its own record, duplicating the information assigned to the "whole pipe"? There is not a single answer to this question; GIS technology generally provides a range of alternatives for dealing with such problems. However, the important thing here is that no alternative addresses each and every possible consideration in an equally optimal way. Also, once having chosen a given alternative, not only custom application development but also data conversion, loading, maintenance, and mapping techniques must be specifically and, perhaps, irrevocably configured around that choice.

- Spatial accuracy: Traditional data tend to be atomic in nature and, therefore, either accurate or inaccurate; a social security number will either match or not match. Some traditional data tend toward a fuzzier interpretation. For example, people's names are legitimately represented in varying ways. Software may have heuristic methods for matching names from a "last, first middle_initial" form to other common but standard representations, but these are not really issues of accuracy so much as data definition. Spatial data, by contrast, must deal with *precision*, which introduces data specification issues that have no common analog in traditional data types. Precision of spatial measurements refers to the range of numeric values that can be stored in a coordinate; for example,

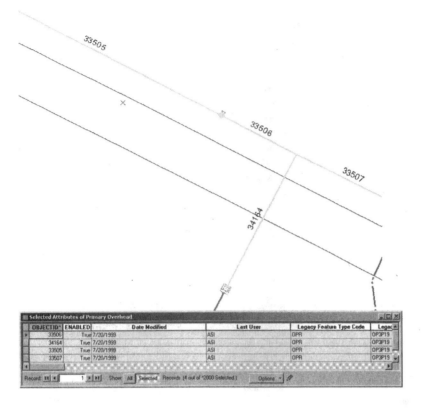

Figure 5.7   Unit definition.

double precision coordinates use eight bytes of data for each dimension (x, y, or z) of a coordinate. While instinct might suggest that higher precision is always better, such is not the case. Along with greater disk space and computational requirements for higher precision numbers, there is the issue of defining when two things are equal: obviously, $4 = 4$ but does $4.0988763421765554 = 4.0988763421766$? The answer to this question cannot vary according to the convenience of the moment. Whatever degree of "closeness" constitutes equivalence determines the minimum spacing between points. Data from different sources or features must be placed within the minimum spacing to be considered connected; conversely, it is not possible to place two separate features any closer than the minimum distance determined by the precision without considering them to be in the same place.

While these special spatial considerations for data definition are critical to determining the potential applications that a given set of data can support, they are just as dependent on the data schema and content.

DATA SCHEMA. The set of all tables, fields, relationships, privileges, constraints, and other software-maintained data properties in a functional portion of a database make up the *schema*. Recalling the earlier statement that applications differ from general software in the way that their actions are specified, we see that the more static data schema are the more automation an application can provide. Extending the example of a tool that sorts by first name, we see that not only is the tool useless in data schema that store all name components in a single field, but also it may be rendered useless if the name of the name field changes. Application developers are constantly responsible for accommodating as many circumstances as can be foreseen, but there is a cost for such flexibility; if, for example, the name of the name field is not coded into the application, then someone else must be responsible for choosing the correct field. This is a step back toward general software, whereas one common reason for custom applications is to automate oft-repeated processes.

On the other hand, application development frequently uncovers new or altered requirements for defining data schema. For example, a given choice for pipe unit specification may have been accepted for use in a project, but when a custom application for exporting the data to modeling software is being coded, the choice may turn out to be unfit. This introduces a chicken-and-egg paradox: one cannot specify the application without first having the schema, but one cannot specify the schema without incorporating the requirements of custom applications. The correct approach is, appropriately, itself a paradox: plan to react. The worst error is to put too much of a limited resource into planning, leaving not enough to react. Of course, failure to plan enough is problematic as well.

DATA CONTENTS. While data schema are of vital importance in defining custom applications, the application developer may realize particular efficiencies, or avoid specific pitfalls, if provided with a truly representative sample set of actual data in addition to the data schema. The advantages of such a "reality check" on delivered application are not likely to be in the realm of what the tool can do, but rather in its performance or "friendliness" to the user. For example, suppose the schema defines a domain for certain pipe attributes (MATERIAL: VC, CI, PVC; SIZE: 2″, 4″, 6″, . . .), but in practice almost always use PVC for 2″. Because other combinations are still possible for both 2″ and for PVC, the application cannot safely set the MATERIAL attribute based on the user's choice of SIZE; however, they could make PVC be the default choice for MATERIAL when the user has chosen the 2″ SIZE. Similarly, complex queries or conditions may be much faster if the most likely data combination is considered first.

Software. Software management is an important and continuous process in a custom application development project. Generally speaking, new releases of software offer enhancements in performance, reliability, and available functionality while attempting to support functionality provided in previous releases. In practice, however, the decision to upgrade base software during or after the implementation and delivery of a custom tool must consider carefully whether advantages outweigh the costs. Some specific considerations are

- New capabilities of the new release may require data, schema, or application changes to exploit; the capability may be available but not applicable in the environment being developed.
- Performance enhancements are potentially the most predictable advantage of a new software release, although the operational effect of performance enhancements is less easy to ascertain.
- Reliability enhancements involve advantages and disadvantages. A custom application developer may have already implemented a solution to bugs in the original software platform but not to new bugs that come with the new release. While formal regression testing may or may not be cost-justified for a new release, it certainly is possible to encounter unexpected consequences.
- Failure to upgrade will eventually lead to an unsatisfying condition; new problems that arise will be ascribed to the "out-of-date" software. The nucleus of application expertise may become less familiar with the "old way" over time and, eventually, the original software platform will be completely unsupported. At that point, it will be harder to migrate the custom application because the original participants may no longer be available.

Hardware. In the context of this discussion, hardware consists of computers and their peripheral equipment (printers, modems, graphic displays, etc.); networks; and specialized equipment, such as GPS units, hand-held data collectors, scanners, and so forth that may be especially important to an application development process. The effects of hardware specification can be far-reaching and fairly subtle. The following are a few examples meant to illustrate how widely diverse hardware issues can be:

- A custom application displays a form on the screen. Certain users are unable to make a selection from a scrolling list. The problem turns out to be that affected users can only achieve a screen resolution of 800×600 pixels, at which resolution the font in the scrolling list is indistinguishable
- Some hand-held data collectors are upgraded. Later on, custom tools loading the data into the production GIS fails with a "datatype unknown" error. The problem turns out to be that the new unit's default date data type is different from the one for which the application was designed.
- A custom plotting application crashes one specific plotter. The problem turns out to be that the custom application requires a font to be downloaded to the plotter that the plotter is not configured to accept.
- A data quality control process begins to run many hours longer than when originally deployed. The problem turns out to be that system backups on another system running over the network are causing a large number of collisions on the affected network switch.

In the above examples, circumstances arose that caused problems in the execution of a custom application; however, solutions to problems were derived outside of the application development group. While it would not be possible to characterize how all such problems arise and are resolved, it is important to note that

- Hardware problems can appear any time something is changed in the entire environment.
- Human nature is such that each unit involved will likely be sure that the problem is somewhere else. This is because people will naturally test what they know best and accept the results as indicating that "their" component is doing just what it is supposed to do.
- Creative problem-solving and excellent communication about and documentation of involved components really pay off.

OPERATIONAL ENVIRONMENT. The operational environment for a custom application encompasses the people, software, equipment, and data that interact to produce a desired result. Each of these is discussed separately in other sections, but certain phenomena are reflective of the ecological principle that a system is greater than the sum of its parts.

One common and often significant impediment to the successful definition and deployment of a custom application is simple human resistance to change; people want the custom application to exactly mirror the workflow that was in place before. A cavalier approach to managing change will not be effective for a variety of reasons, including

- Exhaustively reviewing all of the requirements for successfully completing a certain assigned task is not an easy or immediate chore. If the people conducting that review are hurried, pressured, or disenfranchised by the project team as a whole, errors will likely result.
- One particular situation that is often lost during the requirements review is the handling of exceptions. It is important to explicitly ask the people who are describing an existing workflow about examples of nonroutine decisions that they must make to manage the process that the application will replace. Handling an exception that was not expected may not only cause code errors or unsatisfactory results but, also and perhaps more importantly, the very architecture of the tool may have to be reengineered to respond more gracefully.

On the other hand, organizations can and must respond to changing needs and technologies. An application that can save resources of the organization may, in fact, be devaluing those resources; the individual that used to spend that time is now going to be unoccupied unless and until other activities are assigned. This effect ripples around the organization, reducing management responsibilities of some managers; budgets of some departments; and perhaps even the indirect resources such as personnel, vehicle maintenance, and so forth. If the comprehensive picture of an application's deployment is not understood, it cannot be communicated and if it cannot be communicated, it will likely be resisted.

PERSONNEL ENVIRONMENT. It is vital that a team be assembled that can address all aspects of an application development project. These must include

- Management—managers must ensure that resources are being used effectively to achieve the specific goals required and in a timely manner. It is not necessary that managers take the time to understand all of the technical details involved, but they must understand the "big picture" as well as stay abreast of issues that develop. It is important that managers understand that staff will require much more time to prepare for and respond to planning and status meetings than the meetings themselves require. Especially when a project plan undergoes some iteration, it is easy for managers to underestimate the commitment required by their staff to do a good job of specifying, reviewing, testing, or training on a new application.
- Administrative—administrators provide resources and set the principles on which management decisions are made. It is valuable for an administrator to develop a set of guidelines and measures in nontechnical terms on which to evaluate the goals and progress of an application development process. Typically, measures should be in very broad terms but still narrow enough to clearly discriminate between the application solution and the status quo or an alternative approach.
- Technical—the most critical aspect of application development is that two groups of people, the technical and the operational, learn to speak a common language. Technical people arrive at the problem with knowledge of what the technology itself is inherently capable of. Whenever not directed otherwise, they will approach a problem just the way the underlying software environment dictates. When operational people describe their issues, technical people are hearing the description through a technical filter, perhaps having instinctively substituted things that will work readily for what is really required.
- Operational—on the other hand, operational staff who define the application requirements cannot be expected to master technical language and concepts. One of the most common exchanges heard at the completion of an unsatisfactory custom application is "But it doesn't . . ." followed by "And we told you when we started that it wouldn't." In fact, both sides are telling the truth; the failure is in the different language used by the two sides. The technical people may really have explained a limitation of the technology and the operational people may really have described what they wanted. Solving this communications impasse is one of the more significant challenges facing a project development team.

Certain techniques can facilitate the "language barrier" between technical and operational staff. An experienced, skillful manager overseeing the interaction between them can referee, especially when not otherwise too deeply involved in the process; what is needed is an arbiter, just to make sure that real communication is taking place.

Early training on the operational side is extremely valuable. The more time spent working in the GIS environment, the greater interpretive ability the operational person will have in the technical environment. Typically, it is not necessary in the early stages to train a large group of people, just a critical nucleus who will review all project specifications. Also, note that a common failure in this regard is to arrange for training for the nuclear group without also providing those staff the time, software and data, and motivation to exploit their training. This common failure wastes resources.

Another aspect of facilitating the communications between operational and technical staff is to use vivid and realistic examples as much as possible. It takes more time to illustrate things than to simply describe them in text, but the difference in the effectiveness of the communications is dramatic. On the other hand, one must be careful that everyone is reminded that the example is not the whole story.

Finally, it is advantageous for technical staff to simply observe the target operation first-hand; they may pick up things that are important to the implementation but may not seem so to nontechnical staff. When the application development team is entirely internal to an organization, only management or administrative approval is required to achieve this kind of interaction. For externally produced applications, a modest budget allocation should suffice.

DEVELOPMENT AND TESTING

Development and testing describe the coding and assembly of the various modules defined during the earlier design phase. The development environment should be properly installed and working smoothly on the development machines.

Product development should occur on computer systems that are clearly understood to be "development" computers. In the course of development, many problems can and will occur that may corrupt the operating system, the database, or any number of components in the system. These problems should not affect other systems. In a large organization, a system support team may be available to configure the development environment and should be included in planning the system deployment.

Regardless of the project size, it is important to include all of the elements listed below to maintain the quality required of the final product.

PROJECT TRACKING. Before coding, a method for tracking module assignment, status, and problems should be created. This is especially important if the project is large. However, even if a single developer is assigned to the project, there is value in having these systems in place. A reassignment

---

(even if only "temporary") may lead to a prolonged period of time away from the project. Upon return, there will be time lost to determining the status of the project.

CODING MODULES. Although many programmers enjoy the "artistry" of writing computer code, modern software methodologies, if properly followed, will result in coding tasks that may appear anticlimactic. If the class design is thorough, all that the coding stage offers is to translate the class diagram to the specific platform language. However, it will be inevitable that issues will arise at this point in the process—perhaps a specific assumption or input that was unforeseen previously. It should be expected that designs will undergo review to fix problems and that the management at this point involves solving the "ripple problem", as changes in design manifest themselves throughout the various modules (some of which may already have been completed).

Two important aspects of coding are the proper attention to error handling and coding standards. Attention given to these two areas will provide for greater benefit in the long run because of module reuse and higher quality. The organizational structure of the development team will have an important effect on how well the development is embraced and implemented. Teams with shared capabilities, design reviews, and code reviews will be able to encourage the use of coding standards and leverage previous work that may fit the requirements for some modules. Developers working in solitude can be successful in these areas, but it requires more discipline as well as understanding from management. A developer may realize that the definition of "done" includes proper in-line comments and error-handling capabilities. The short-term pressure to complete features for the product will conflict with these long-term goals.

WRITING A TEST PLAN. A test plan is required to properly evaluate the application and determine when modules are, indeed, "done". The elements of a test plan should include the project members who will actually perform the tests. It is strongly suggested that project members who code the modules not perform the tests on those same modules. Elements of a test plan will also include

- Objectives and scope of the test;
- Use of automated scripts, user interaction, or manual checklists;
- Listing of test data required;
- Required "environment" for testing (number of users, network traffic, etc.); and
- Detailed test scenarios, including expected results.

Some commercial packages are available to help write and execute a test plan. Some are specific to a development environment and can be used to track specific lines of the program that are executed during each test (this is referred to as "coverage", not to be confused with a specific GIS data format).

The test plan should be considered to be a living document that will be updated frequently. It should contain the references to test scripts or manual checklists used for unit, integration, and system testing. The record of tests (when they were run, who ran them, the results, etc.) may be included as part of the plan or kept as a separate project document.

TEST DATA. The test plan should discuss the kind of data required to sufficiently test the software. In environments in which data are changing during the course of running the application, this may need to be simulated by writing another program solely for the effect of updating particular tables (i.e., simulating the dynamic nature of the database). If processes outside of the scope of the application are responsible for the integrity of the database, database faults should be tested to see how the application behaves. In addition, for performance testing, the size of databases should approximate those being expected during the production phase of the application.

UNIT TESTING. A unit test is performed on the smallest (or lowest level) piece of coding that can be determined to be a module. Based on the module's specifications, a test program is written to pass both proper and improper inputs to the module, and the results are then observed. It is important to ensure that all sections of the code are executed during the test (see the reference to "coverage" above).

INTEGRATION TESTING. Unlike module, or unit testing, integration testing occurs as part of the integration process. Units that have passed the unit-testing phase are assembled into larger components. Rather than combine every module and attempt to integrate the entire application at once, it is suggested that test "stubs" be written that contain a minimal amount of code but that serve as place-holders for modules that have not yet been tested. In this manner, two units can be integrated and tested. As each integration succeeds, successively more test stubs are replaced by true modules. At each stage, integration tests are rerun. Results of the integration test will change as more modules are added. These results should be saved (either screen captures, text output directed to files, or database results) at each stage of the test. Future testing required as part of regression testing should compare the results of the tests against the expected results. Regression testing refers to the concept of reproducing results at each stage of testing. Because changes often occur in requirements (for any number of reasons), each stage of testing should be rerun to determine any unanticipated effects to the changed code. For example, external software may be upgraded to a new version. In this case, the application will need to be tested with the new environment, and any failures will need to be identified through use of the test plan and comparison to its previous results.

SYSTEM TESTING. System tests are performed when integration testing has succeeded using all of the defined modules (also called a complete "build" of the application). All aspects of program interaction should be included, such as logon procedures (e.g., how logon failures are handled); help file interactions, if applicable; and resource interruptions (e.g., network).

USER ACCEPTANCE TESTING. This phase (sometimes called *validation testing*) aims to demonstrate that the software functions in a manner that can be reasonably expected by the customer. Under some conditions, a client will require a formal acceptance test. This will clearly document the completion of the requirements as interpreted by a real user of the program. This should resemble acceptance tests performed in other industries. A list of unresolved issues (a "punchlist") may be required and a timetable should be established for resolution of problems.

If software is being developed for use by many customers, it is often unreasonable to perform acceptance testing for each customer. An alternative is to use selected customers at late stages of the development process to uncover previously undetected errors. There are often two phases of user tests, referred to as alpha and beta. Alpha testing is conducted in a controlled environment, typically with the developer present noting each mouse click, etc. Beta testing is not strictly controlled, is typically conducted at multiple sites, and requires users to report problems back to the developer.

*D*OCUMENTATION AND *TRAINING*

COMPONENTS OF DOCUMENTATION AND TRAINING.
Documentation and training overlap significantly in their content, organization, and responsible source, but they are two different things. Documentation is a resource used by a person whose agenda is self-determined; it may be an end user trying to understand which button to click, a systems engineer trying to understand the effect of the application on the network, or perhaps a manager trying to determine whether it will be possible to alter a workflow for which the application is just a part.

Training, on the other hand, is a directed activity; there may or may not be an instructor but something or someone coordinates a trainee's activities and environment to achieve a specific sequence of events in which the trainee sees something like their target results when executing something like their required action. The results are similar to physical training, where repetition leads to greater facility and comfort with the activity.

Documentation is relatively time-independent; the materials may be read or reviewed at any time and in any order. The time for training deserves significant consideration. Rolling out a new application without prior training is often catastrophic, yielding a level of frustration that may never be surmounted.

On the other hand, training too early can be quite ineffective as well. For one thing, the revisions of function and scope that often accompany the development of a complex system may fundamentally change some aspects of the application's operation; user's trained the "old" way will have to be trained again. Equally important, training is really only an introduction; if a trainee does not quickly follow up formal training with a period of real experience with the tools, he or she will forget what was previously learned.

Documentation may consist of formal electronic or hard copy documents, online help or pop-up messages, memoranda, e-mail or meeting notes, or even dynamic Web sites that are (theoretically) up-to-date. Documentation is almost always the responsibility of the developer; however, it is wise for the recipient to allocate resources to review submitted documentation for clarity, completeness, accuracy, and style.

Training is often developed in a cooperative manner between the developer organization and the client organization. An initial training module or modules may be sketched out and provided by developers for a small nucleus of senior staff on the client side. This training will serve both to evaluate the effectiveness of the training approach and prepare the nuclear group for disseminating the information internally.

One specific consideration about training and documentation for GIS applications is that, almost always, the custom application is developed in the context of a software package that is itself valuable, expensive, and fully documented. It will not be often that a client organization can be proficient with managing a custom application while not understanding the underlying software.

Technical Resources Used in Application Development. Too many organizations allow documentation and training materials to drift away from the evolved product. Avoiding this common pitfall requires a scrupulous adherence to a simple principle: if you change what it does, change what you say it does. Achieving this simple result will require the habit of exchanging document versions, perhaps quite often. A word-processing environment common to all participants and facility with the use of version management or reviewing tools in such software will be helpful. The time required to review documentation under development in a somewhat broader context than just the application developers and specifiers is one of the best ways to inform administrators, managers, and end users of what lies ahead, although it is not appropriate for every step in the development process.

Documentation Created as a Byproduct of Application Team Intercommunications. Developing a central repository where all meeting minutes, memoranda, e-mail exchanges, and so forth will be collected and made available is valuable in itself and also aids the documentation development process. Along with clarifying details of the technical requirements, this body of information demonstrates the level and style of language with which participants are most comfortable. Often, the copying of text from an early specifications document to a final product help page actually makes it easier for readers to understand the big picture; they understood what it meant when they asked for it and now they see where it fits in the delivered tool. Of course, literally copying text has its pitfalls, too, which is why review is important.

Internal Programmer Documentation. While requiring or reviewing the documentation that programmer's put in their actual code may be both impossible and irrelevant for the client organization, there may be some value to discussing the topic with a prospective consultant. An organization that is committed to producing good software tools and supporting them well over a long period of time will have a process for ensuring that programming techniques are fairly standard and methodical, allowing any available professional employee to jump in and understand how the code is working. Lack of a formal process for code maintenance does not prove that you will not be able to get anyone to help you further down the road, but the presence of such a process implies that you will.

In some environments, the client organization will be receiving code in a readable form, perhaps with the specific purpose of maintaining the code itself over time. In this case, good programmer-level documentation is an absolute necessity. Furthermore, the overall documentation for the product should include a programmer-level description of the internal design of the code.

Online User Documentation. Online documentation may include help information that is available in the context of the specific task in which a user is engaged. Where there is a button to click, there can be a help page to explain what is supposed to happen when the button is clicked. This approach to documentation is fairly expensive to build and maintain and most certainly cannot be added easily later. For a critical application in which large numbers of relatively untrained staff are going to be using the custom tool, the cost of context-sensitive help may be justified. Another consideration with these types of materials is that clarity or appropriateness of the information that pops up is user specific. It will not be possible to specify simply that "Good, clear pop-up help will be available for all application functions". Rather, a significant effort on the part of the client will be required.

Training Materials. It is useful to have training materials that are readily copied, with supporting data that reasonably represent what users will actually be dealing with. Often enough, a trainee will have an application problem that seems similar to one dealt with in training. If the user can readily get to that exact step in the training sequence, the user may be able to educate himself or herself from the training materials. Also, a trainee armed with the training materials at the individual workstation can follow up the training with repetition and experimentation, which is helpful for development.

PERSONNEL INVOLVEMENT. Allocating personnel to the documentation and training requirements of a custom application project is easy to underfit. Poor documentation and training scuttle difficult projects and at least retard projects that flow easily to rollout. It is best to use a comprehensive approach in which documentation and training are counted as driving forces from the beginning, while at the same time causing those products to "congeal" along with the specifications and implementation. Achieving this goal requires a team approach in which roles, mechanisms, and management milestones are properly defined.

Operational Staff. Operational staff, meaning those who work with the business processes that the application enhances or with which it interfaces, are obviously critical to the development of training and documentation materials. Application interfaces that naturally reflect operational sequences in use generally require less documentation. Good documentation and training materials are expressed in the conceptual framework of the people who use them. While the actual creation of documentation and training materials may fall to other groups, operational staff provide, by far, the most important review. If they cannot understand it, they cannot use it, and if they cannot use it, it does not work.

Technical Staff. Technical staff, meaning those who either specify or implement the custom code, must be encouraged to look at the development of documentation and training materials as being on their critical path to success. Assure that these staff have time to work on the materials and that the value of their time is validated by committing to a real review process.

PARADIGMS OF DOCUMENTATION AND TRAINING. There is an abundance of literature on the subject of training philosophy, far beyond the scope of this writing. Much less is written about the subject of software documentation except by the sellers (who claim that it is good) and the buyers (who claim otherwise.) Two specific paradigms are mentioned here because they emphasize the holistic approach, a lack of which is often destructive to GIS projects.

Documentation Before Development. As computer software has grown more and more intelligent, not merely blazing through rote processes but actually doing something like thinking, the dream has periodically arisen of simply telling the computer what is desired, and letting it figure it all out by itself. While such a thing has never really happened, one can benefit from taking that first step: write down exactly what is desired in the custom application and let that serve as the specification. People are invigorated by pictures, examples, and working details—so is documentation. Ultimately, nearly every statement describing how a tool works expresses a decision made by the specifier and communicated to the implementer.

Nuclear Dissemination of Information. Anyone who has ever been frustrated by computer software that does not do what they want will agree that it would be better to have a person to ask than to flip repeatedly through the same pages of a manual that always seems to be talking about something else. Nuclear dissemination of information is a paradigm that says that training dollars should include a higher level of training for a small group of people distributed through the organization—they become the "go-to" folks. Not only does that save a lot of time and effort but, because they are accessible to the users at the moment they are needed, they can raise the level of understanding of their colleagues exactly where it is needed.

*R*EFERENCES

Alston, R.; Donelan, D. (1993) Weighing the Benefits of GIS. *Am. City & County*, **108** (11), 14.

Barnes, S. (2001) ESRI and Sun Launch LBS Portal. *Geospatial Solutions*, **Feb**, 16.

Boehm, B. W. (1988) A Spiral Model of Software Development and Enhancement. *IEEE Computer*, **21** (5), 61–72.

Booch, G. (1994) *Object-Oriented Analysis and Design with Applications,* 2nd ed.; Addison-Wesley: Reading, Massachusetts.

Coad, P.; Yourdon, E. (1991) *Object-Oriented Analysis,* 2nd ed.; Prentice Hall: New York.

Engelhardt, J. (2001a) GITA's Technology Report. *Geospatial Solutions*, **Feb**, 14.

Engelhardt, J. (2001b) Mapping the Future. *Imaging Notes;* Space Imaging: Thornton, Colorado; May/June, p 30.

Estes-Smargiassi, S. (1998) Massachusetts Water Resources Authority Uses GIS to Meet Objectives Cost-Effectively. *Water Writes;* Environmental Systems Research Institute: Redlands, California.

Farkas, A. L.; Berkowitz, J. B. (2001) State-of-the-Industry Report. *Environ. Eng.*, **37** (4), 22–26.

Geospatial Information and Technology Association (2001) Geospatial Technology Report; www.gita.org (accessed Jan 2001).

Geospatial Solutions (2001) Innovative, Imaginative, Intricate—Top Apps 2001; **Aug**, 16.

Kruchten, P. (1999) *The Rational Unified Process: An Introduction,* 2nd ed.; Addison-Wesley: Reading, Massachusetts.

Lanfear, K. J. (2000) The Future of GIS and Water Resources. *Water Resources Impact*, **2** (5), 9–11.

Monmonier, M. (1996) *How to Lie with Maps*, 2nd ed.; University of Chicago Press: Chicago, Illinios.

Royce, W. (1998) *Software Project Management: A Unified Framework*, 1st ed.; Addison-Wesley: Reading, Massachusetts.

The San Diego Union-Tribune (1998) GIS–New Technology's Scope is Almost Infinite. Jul 26, 1998; p A-12.

U.S. Environmental Protection Agency (2000) *Environmental Planning for Communities—A Guide to the Environmental Visioning Process Utilizing a Geographic Information System (GIS).* EPA-625/R-98-003; Technology Transfer and Support Division; Office of Research and Development: Cincinnati, Ohio.

Suggested Readings

Frank, A. (2001) Building Better Maps. *Imaging Notes;* Space Imaging: Thornton, Colorado; May/June, pp 18–19.

Heaney, J. P.; Sample, D.; Wright, L. (1999) *Geographical Information Systems, Decisions Support Systems, and Urban Stormwater Management.* Cooperative Agreement Rep no. CZ826256-01-0; U.S. Environmental Protection Agency: Edison, New Jersey.

Huber, W. C.; Dickinson, R. E. (1988) *Storm Water Management Model;* User's Manual; Version 4; Environmental Research Laboratory; Office of Research and Development; U.S. Environmental Protection Agency: Athens, Georgia.

Jenkins, D. (2002) Making the Leap to ArcView 8.1. *Geospatial Solutions,* **Jan**, 46–48.

Robertson, J. R. (2001) Feeding The Flames—Airborne Imagery Fuels GIS. *GeoWorld,* **14** (3), 36–38.

Shamsi, U. M. (1998) ArcView Applications in SWMM Modeling. In *Advances in Modeling the Management of Stormwater Impacts;* James, W., Ed.; Vol 6; Computational Hydraulics International: Guelph, Ontario, Canada; pp 219–233; Chapter 11.

Shamsi, U. M. (1999) GIS and Water Resources Modeling: State-of-the-Art. In *New Applications in Modeling Urban Water Systems;* James, W., Ed.; Computational Hydraulics International: Guelph, Ontario, Canada; pp 93–108; Chapter 5.

Shamsi, U. M. (2002) *GIS Tools for Water, Wastewater, and Stormwater Systems;* American Society of Civil Engineers: Reston, Virginia.

Turner, A. K. (2001) The Future Looks Bright (In Any Spectrum) for GIS Data. *GeoWorld,* **14** (5), 30–31.

Chapter 6
Geographic Information Systems Data, Software, and Hardware Maintenance

INTRODUCTION

One of the most significant issues that affects the reliability and accuracy of a geographic information systems (GIS) application is its scheduled maintenance. Currently, the GIS user community requires extensive metadata (data about data) that describes where and when the data was created. Because of this challenge, it is necessary to have the most current version of hardware and software to perform daily and vital GIS operations. Being able to carry out these tasks in a timely manner and produce up-to-date and accurate datasets enables a GIS to be reliable in today's information technology society.

Having a GIS does not only include the hardware and software for data processing and analyses. The system should also incorporate a set maintenance schedule for the datasets that are integrated within it, because the system is only as good as the data it contains. The effect of using out-of-date data is cumulative with time. New information is continually emerging and modifications are taking place, which could make the existing databases obsolete if they are not properly maintained. Furthermore, the world is dynamic and so must be the georeferenced features that are used to track it.

Decisions based on GIS data increasingly affect daily activities. The moment a parcel is divided, a new street put in, or a utility pipeline replaced, the database needs to be updated to keep it current. Once the data has been updated, it can be used by various GIS users in their daily tasks, i.e., emergency management, fire fighting, disease monitoring, and construction management.

To keep the GIS data current and provide an accurate basis for analysis and decision-making, GIS data should be updated regularly. Support on behalf of the system's users (developers and end-users) is required to accomplish this task. To be successful in achieving these GIS services and retaining effectiveness towards the decision making process, implementation of a scheduled maintenance program is essential.

Therefore, GIS maintenance includes a series of corrective revisions and required expansions. These operations are used to maintain the functionality of the database while preserving its integrity. Therefore, the GIS team that is overseeing the integrity of the system should keep a regular maintenance schedule and document changes in the datasets to sustain a reliable system for all users.

BACKGROUND. A computer-based GIS can be used to assemble, store, manipulate, and display geographically referenced information. For some applications, such as fire and police emergency response, hydrant location, sewer and storm drain alignments, it is important to have the most current and accurate data available. For other purposes, such as planning, zoning, and property assessment, where data changes more incrementally, the frequency of updating can be relaxed.

A substantial amount of time may pass between the needs-assessment phase and final delivery of the GIS product, software, and hardware. Many changes and modifications can occur, because GIS is dynamic and the technology is continually advancing. The primary GIS components (the computer, hardware, software, and data) are constantly changing with advances in technology and methods used for data collection and conversion. Careful steps should be taken to ensure that delivered components meet project specifications for final acceptance of the datasets. This will enable the maintenance personnel to certify the proper data are compatible, and an ample evaluation period is available before final acceptance to assure the functionality of the system.

A well-documented, deliverable product based on proven dataset standards, database integrity, and GIS procedures can overcome drawbacks that exist before the system's release.

Necessary steps to implement a cost-effective GIS maintenance program include verification of the data source, assessment of data accuracy, and selection of available software and personnel. Some of the causes leading to the need to maintain the GIS datasets are long conversion time periods, missing source data during conversion, and a continuing production of new data that must be integrated to the GIS to have a complete and current dataset.

DATA. Carefully maintained and regularly updated data is the most valuable component of a GIS project. The dataset is comprised of spatial features such as points, lines, polygons, and relational data. Relational data includes attributes and textual information that describes the spatial features. The primary reference in a spatial database represents a defined coordinate pair (x, y) in geographic space, i.e., latitude and longitude.

Data can come from various sources and can be in various formats. In these cases, data-related activities such as data acquisition, conversion, updates, interpretation, and maintenance costs must be taken into consideration during the GIS project. Planning costs can be minimized through the use of the following data maintenance strategies: data integrity, data sources, and leading cause. However, the key is to get the data transferred to the GIS database to reflect real world conditions.

Data Integrity. Geographic information system databases should always be consistent and up-to-date. Every change or update to the database should be done through a structured, systematic process. To achieve this, it is recommended to use a primary consultant to input the spatial and attribute data. Establishing a maintenance and update process requires a set of policies that the consultant should follow when updating the database. A well-defined approach can make the process efficient and cost-effective and protect the integrity of the data.

Data Sources. Geographic information system base data can come from multiple sources. These sources may include as-built drawings, orthophotos, parcel maps, and utility plans. An influential factor that must be considered for the GIS maintenance program to function is the source of each dataset. An efficient way to input and update the GIS database is to extract data from as-built drawings. In a GIS, it is also practical to link the scanned images of as-built drawings to the related features, providing additional access to the original information.

Leading Cause. Some of the challenges with the receipt of GIS datasets that can lead to initial maintenance issues are

(1) Schedule delays. Schedule delays occur because of failing to follow the set production schedule. To eliminate this problem, project management should meet regularly to set a schedule for dataset delivery. With regular meetings and correspondence, the initial schedule can be amended based on constantly changing demands within the project. Because of the amended schedule and regular meetings, the project can continually move forward and address issues as they arise without compromising the stability of the project.

(2) Inadequate conversions. Inadequate conversions are errors carried out by either the selected consultants or the conversion group. To eliminate this problem, key personnel should be trained in the conversion process that will be implemented during the project.

(3) Alteration of the software and hardware during the conversion process. Data builds up because of the lack of standardized conversion programs and hardware to run the programs. To eliminate this problem, the conversion process should be tested during the pilot phase of the project. The conversion personnel involved and the management team can see the changes that must occur to the software and hardware before full production of the data.

(4) Data that does not conform to the database design specifications. This causes the creation of a backlog because of the need to perform extra conversion steps on the data. To eliminate this problem, incorporate data during the pilot phase that does not conform to the database design. With this incorporation, the conversion personnel will be able to alter the software to recognize common problems that may occur in the datasets when the project moves out of the pilot phase and into full production. Therefore, the conversion software will be able to show personnel data discrepancies in a timely manner so that software fixes can occur quickly and without holding up production.

(5) Inability to meet data posting demands. This causes the creation of a backlog because of the need to deliver datasets on time. To eliminate this problem, management should meet during the pilot phase to set time-sensitive (weekly, monthly, quarterly, etc.) goals for data delivery. Once the goals have been set and regular meetings are held to monitor the project's progress, updates to the goals can be made to satisfy all members of the project, including the client.

EFFECTS. Proper maintenance is a critical element in the success of a GIS database. Lack of planning towards the maintenance and scheduled updates to the database can result in the demise of the GIS project. A fully operational

GIS not only requires a well-maintained system of spatial data, but also necessitates regular upgrades for software applications and hardware technology.

URGENCY. As the amount of newly acquired data grows larger, GIS data can become inaccurate and subject to loss because of not following the procedures set in the pilot phase of the project for data delivery verification. Because recent data is often in the greatest demand, unregulated data can cause the GIS to appear limited or inaccurate. These backlogs should be cleared as soon as possible and the database should be updated accordingly. By following a set procedure for data delivery verification (which includes monitoring data delivery within the project's database), data backlogs can be avoided.

PLANNING AND STRATEGY. A proper maintenance plan is crucial if the GIS is to remain useful for future projects and users. Careful planning is a critical element for successfully undertaking a data maintenance project. Furthermore, GIS data sharing is an important aspect of a successful GIS system. When different users are involved, the participants, depending on their needs in the organization, should negotiate to schedule the implementation of a data maintenance plan.

In a city, participants typically include utilities, public works, fire, police, and planning. The needs of these different participants determine the data and mapping usages for the GIS. The needs of each of these groups should be addressed in a well-defined approach. This approach should take the form of meetings that focus on needs-assessment discussions for each of the participants. The outcome of these meetings should be a needs-assessment document that gives specific requirements from each of the participants for the GIS's purpose. Examples of what is required from the GIS are regular updates, feature and database maintenance, and usability of the system by specific types of users, i.e., novice to expert users. A combined approach of planning and implementing an effective needs-assessment plan will help the GIS to remain cost-effective as the project matures.

TYPES OF MAINTENANCE. The various types of data maintenance can be classified as

- Relational data or attribute updates. Updates in this category are those that amend the properties of the spatial features in the database.
- Spatial data or graphical updates. This type of maintenance focuses on the revision of the graphical features that represent each record in the database.
- Preexisting backlogs. These are data that were added subsequent to preparation of the original GIS database or when the conversion program was carried out.
- Incidental data updates. These changes come online whenever there is a single addition or modification to the GIS database.

- Schedule-derived delays. These datasets result from delays in the planned GIS conversion schedule.
- Ongoing or regularly scheduled updates. These changes consist of regular updates, such as land base improvements or subdivisions of parcels.
- Backlogs because of a lack of resources. In this case, data are accumulated because there are not enough resources or personnel available to keep the data current. Some of the backlogs may be generated because of the capacity of a particular hardware/software configuration or because of a lack of personnel.

SIGNIFICANCE. Up-to-date datasets are an integral part of any GIS. While personnel involved in some traditional analyses can accept data which may be several months old, emergency services provided by most utilities require up-to-date information reflecting the latest changes in the field. Therefore, because of this need for current datasets, a comprehensive maintenance procedure is required to improve data accuracy and analyses and decrease production time for each dataset.

DATA ACCURACY AND CONFIDENCE LEVEL. Data accuracy is a primary requirement for any GIS database. A GIS database is typically based on static engineering drawings for utilities or any relevant sources based on its use. These drawings are typically not suitable for interpretation by GIS technicians and extra caution should be paid to those who extract the data during the conversion process. Geographic information systems data created during this step are typically used both for mapping and analyses. Any data used for modeling and analyses require the utmost accuracy in the datasets.

During the update process, data accuracy should be monitored carefully for any discrepancies that could corrupt the integrity of the GIS database. Monitoring the data for discrepancies includes both spatial accuracy and feature attribution checks. The positional accuracy of GIS data depends on the accuracy of the base map. Base maps are created either through a surveying or GPS process, or they are extracted from aerial photography. Horizontal accuracy for maps extracted from aerial photographs depends on the resolution. With a large-scale-resolution aerial photograph, a technician will be able to extract more features on the ground because of the level of image detail. Use of digital orthophotos can be both a less expensive and more attractive option for building a GIS land base map system. Orthophotos can be a useful source for extracting various utility features to maintain and update the GIS database.

Data accuracy should be the prime consideration for GIS database development from the beginning, not just through the maintenance stage. Experience with accurate data gives users confidence when making decisions that otherwise would have to be supported with inaccurate data and varying levels of accuracy.

ERROR REPORTING. Correcting data inconsistencies is an important step in the maintenance process. A standardized plan should be developed by the responsible staff to document the identification of inconsistencies in the data for proper maintenance of the GIS. Once the plan has been documented and approved, the GIS maintenance staff should correct the data and document which updates were made to the database and datasets in the corresponding metadata. Errors that have been identified by maintenance personnel should be corrected in a timely manner and the database should reflect these changes when they occur.

RESOLUTIONS. The first step towards resolving data accumulation is understanding the multiple aspects of data maintenance for GIS datasets. Then, after identifying the source of backlogs, the issues that created the backlogs should be addressed by following specific maintenance procedures and educating personnel to help solve the issues associated with data maintenance. Therefore, to address these issues, a procedural document should be established.

A backlog often develops during the period between conversion and implementation and testing. It is often the case that certain datasets were missed or the decision made to add them during the update process, which contributes to the potential backlog. To decrease the likelihood of missing data while updating the database, data input should be prioritized and scheduled accordingly. Manual input backlogs should be eliminated by using automated procedures.

UPDATE PROCEDURES. Procedures should be set during the pilot phase of the project that will enable the GIS (maintenance) staff to keep the GIS and its data up to date throughout the project. The initial efforts of this plan should be focused on backing up the data on a regular basis (i.e., daily, weekly, or biweekly). Backup of the data can take the form of CD or DVD archives or storage on a separate server that is designated as an archival-only server.

Along with archiving data on a regular basis, data should be maintained to keep it accurate and up to date. This task can be achieved through routine updates and modifications to the datasets. By setting a regular update schedule and allotting some time for modifying the data to adhere to changing demands in the project or project data structure, the GIS can be maintained to meet project and client specifications.

Furthermore, expansions to the GIS will enable the project to encompass new aspects that were not planned for during the pilot phase of the project. Such expansions, i.e., adding new types of utilities to a GIS utility project, can not only fulfill the requirements of the project, but also allow the project staff to deliver a product that is more useful on a long-term basis. However, it should be noted that adding new, unplanned data could be a hazardous venture because it can lead to new types of problems. Therefore, if expansions are added to the GIS, they should be done in moderation to create a more complete product, as opposed to a new problem.

With any update process, new software and hardware are incorporated. The software portion of the process should incorporate updates to the existing vendor software (i.e., Microsoft Office, ESRI ArcGIS, and Oracle or Novell Server Updates) and applications that have been developed in house. Updates to the existing software should be incorporated, but with some restrictions. Such restrictions should focus on the need for the updates and what new functionality the updates will add to the existing software. On the other hand, hardware updates should not be as frequent as updates that occur to the software because of cost and necessity. Because of the initial costs of hardware, which can be quite high, preparations should be taken during the pilot phase of the project to update any hardware that doesn't meet the requirements of the project. Hence, if a hard drive on a computer breaks during the course of the project, (minor) funds can be allocated to replace the hard drive and not the entire computer (major funds).

Also, data input methods should be addressed in the early stages of the update process to find out what will work best for the project. Whether the data is updated directly through the GIS software or through automated scripts, the input methodology should be tested and then implemented once a procedure has been agreed on.

Archiving data on a regular basis enables the maintenance team to continually move forward with the project because the team will avoid discrepancies that can occur with server or software crashes. Moreover, data that is kept up to date allows the team to compare data for levels of accuracy. If a comparison to the acceptable data shows that the newer data is not as accurate, then the newer data can be updated to the level of accuracy that meets the project's standards.

As the project moves forward and the GIS is maintained, quality control should be implemented. Practices focusing on visual checks and running automated procedures to check the data's accuracy should be implemented to meet the data specifications for the project. These methods can vary from project to project and should be addressed during the pilot phase to specify what methodology should be used for the project at hand.

Once all of the datasets have been created and kept up to date and accurate, data distribution is implemented. Distribution of the data can take various forms, depending on the needs of the clients. The finalized data that is sent back to the client can be sent through e-mail, on CD or DVD, or through paper reports and map sheets. The delivery method should be set by the client in the initial phase of the project, and the data's quality should not vary during the delivery process. To verify that the data is not altered during the delivery of the data, documentation should be written to establish the quality of the delivery package before, during, and after the data is sent to the client. All personnel responsible for the delivery of the data should sign the documents, and copies should be archived for legal purposes.

As a final point, archives of the data used during the project should be kept. This data can be used in future projects that can benefit from recent datasets. Moreover, because of the data being up to date, newer procedures and programs can be written that are based on the data. For example, if computer-aided design centerline data is the result of a recent update project, then it can be used to write a program that updates the E-911 street centerline addresses that correspond to the latest street grid within the project area (i.e., neighborhood, town, city, or state).

RESOURCES AND COST. An important factor in maintaining a GIS database is procuring trained personnel with specific skills. Staffing issues can be addressed by carefully evaluating the current strength and skills of the staff involved in maintaining the GIS. Based on the anticipated GIS data needs and confidence level with existing datasets available, resource availability for staffing can be planned and implemented. Criteria for selection of data maintenance levels should be included during interdepartmental resource availability assessment. Clearly defined responsibilities, funding source identification, participating group needs, and skills required for the critical tasks should also be taken into consideration during this process. Once the staff is retained, comprehensive data maintenance responsibilities should be clearly defined and established with the specific tasks of updating and maintaining the data.

Budget allocation greatly affects the collection of mapping features and attributes that are included in the database design. Ultimately, GIS data maintenance frequently results in greater costs compared with initial database development. As with data collection, maintenance costs increase with the accuracy to which the data is maintained. The higher the requirements for data accuracy, the more costly the data will be to convert and maintain. In a multidepartmental organization, accuracy standards should be established during the GIS planning stage.

It is crucial to have skilled personnel handling data maintenance, otherwise the chances of having unreliable datasets will be high and the system will most likely be corrupted. This situation can be solved by providing the maintenance staff with proper training to improve the skills of the personnel who are managing and maintaining the GIS data.

MAINTENANCE RESPONSIBILITY. Maintenance of the GIS and its database requires even more attention than initial data collection. When the GIS database is created, it is common for untrained personnel to create a database that contains multiple errors. These errors can affect the overall structure of the database, and even cause future updates to be inconsistent. Thus, these mistakes and missing information must be corrected and integrated to increase the confidence level of the database. Most GIS databases are very dynamic, changing almost daily with new application demands. Formal procedures for comprehensive maintenance and updating activities must be created, documented, and followed-up by all GIS staff and end-users. These procedures will enable the GIS to be more successful and confidently relied on than a system that does not have a procedural document to standardize its use.

In a multidepartmental organization, GIS maintenance and update responsibilities should be thoughtfully assigned with accountability organized to assure that data remains accurate and up to date. Irregular updates can result in the loss of GIS data credibility and essentially render the database unusable.

ADVANTAGES. A GIS will be effective and trusted only when current and updated data is used. Software upgrades enhance performance and keep the application current with the latest technology. Regular upgrades to the application keep the software in compliance with the latest technology and must be a part of the ongoing expense in the project's operating budget.

A well maintained and reliable GIS dataset inspires confidence in users. It speeds data analyses through an organized dataset structure and enables personnel to produce publication-quality maps without dataset quality or accuracy issues.

MAINTENANCE PROGRAM IMPLEMENTATION. There are two ways to manage maintenance updates of the GIS. The first is through in-house standards and procedures that regulate when and which updates occur to the GIS and its databases. By updating the GIS data in-house, set procedures that are already in place and formalized do not have to be changed to fit the needs and methods of other consulting firms that might have otherwise prepared the updates. However, as workloads increase and updates become more frequent, the need to follow the second method of update maintenance becomes more prevalent. Outside assistance is typically more effective and can easily be obtained for acquisition or modifications of data-related tasks. Outside assistance is typically more effective and can easily be obtained for acquisition or modifications of data-related tasks.

The second method to manage updates to the GIS is through outsourcing. By outsourcing updates to a capable firm, agency GIS staff can focus elsewhere and more work can be accomplished in a shorter time span. However, outsourcing would involve regular, procedural checks on the work that is being done by the consultants. To neglect or minimize the requirements typically has negative effects on the workflow of the update process, risks loss of procedural documentation, and may result in the consultant producing an inaccurate product.

Both methods have positive and negative aspects. However, project workflow should be observed and documented during the pilot phase to make a more informed decision when implementing the update maintenance program.

SOFTWARE MAINTENANCE

In recent years, GIS software and hardware functionality has expanded tremendously. Keeping pace with new releases can be cumbersome. A good policy to follow when planning and purchasing new software and hardware for a specific GIS is to include a maintenance-and-update program with the potential vendors. This will enable the GIS to stay current with the rapid growth experienced by this industry.

The chosen GIS software should provide a good integration of attributes and graphics, maintain data integrity, and have a reputation for reliability. It should be flexible to accommodate the changes and modifications required for the system. The software should also serve the user's needs and function as a comprehensive, multiuser tool for GIS analyses. It is important to use the vendor-offered software update program to keep the latest version of the software available to staff.

HARDWARE MAINTENANCE

It has historically been necessary to frequently upgrade the project's hardware to keep up with changes in technology. A GIS requires the handling of mass datasets and complex linkages, making it crucial to choose a faster processing speed computer system with a large memory. By cutting costs in the budget for computers, the project can suffer. This strain can take the form of computers that cannot perform the daily tasks of the project. Hence, in-house personnel lack faster and more complete analyses and productivity time.

Faster data analyses results and access are important factors in managing GIS databases. Although the experience of hardware and software enhancements is not always pleasant and newer versions of the software may involve changes to the data structure, migration to newer database formats and spatial topology can assist a company in being better prepared with newer capabilities and faster processing computers. Speed is necessary in various graphic operations, such as map loading, map display, plotting, and analyses. However, it should be noted that if the current GIS project calls for computing capabilities that are currently available in-house, then technology upgrades are not necessary. Thus, only if the technology limits the company's processing and analyses ability should additional and/or newer computers and software be purchased. As GIS databases continue to expand in size and complexity, faster access to datasets has arisen as an important issue for the industry.

The cost of regular system updates and maintenance should be justified by the specifications set forth in the project's contract. If the current, in-house technology does not meet the requirements of the project, then updates are necessary. The benefits that are associated with newer technology (less maintenance time and faster processing capabilities, etc.) will exceed the incremental productivity costs needed. Again, a review of the current technology and scope of work for the current project should be performed to weigh the needs of the updates.

SUMMARY

It has historically been necessary to frequently upgrade the computer and hardware to keep pace with technology. A GIS requires the handling of mass datasets and complex linkages, making it crucial to choose a computer system with a faster processing speed and a large memory. The cost savings from more economical systems are typically less many times over by limiting staff productivity.

The inclusion of GIS in today's information technology-based society has experienced various setbacks and conflicts because of evolving data formats, format extinction, and lack of planning with the newer data formats. Hardware and software vendors that remain committed to GIS have been vigorously updating their products to keep up with the needs of the GIS user community.

The evolving needs of the GIS community have persuaded key decision makers to enforce more procedural documentation when working on GIS projects. Through staff and management efforts and enforcement of documentation preparation, procedural documentation has become an essential component for effective GIS maintenance.

Current GIS hardware and software have enabled new and existing users to glimpse the technology that will lead them into the future. This technology, which considers the needs of the GIS community more than in the past, is easier to use and is more effective through enhanced, analytical capabilities.

Most GIS datasets comprise a large amount of data, resulting in the necessity for high-end data storage. In addition to managing the data and keeping the data current, a state-of-the-art hardware and software system is recommended to minimize the display and analyses time.

The evolution of GIS is helping the world to improve monitoring and analyses efforts ranging from sewer pipeline quality to disease monitoring. With the aid of modern GIS maintenance programs, this evolution will continue to regulate the effectiveness and efficiency for all data within a GIS.

*S*UGGESTED READINGS

Advanced Geographic Information Systems (1998) http://www.valpo.edu/geomet/geo/courses/geo415/projmgmt.htm (accessed March 2001).

AGRA Baymont. Data Maintenance and Backlog in AM/FM/GIS Systems. http://www.baymont.com (accessed March 2001).

AXON's Data Tracker (1999) Data Tracker GIS. http://www.theaxongroup.com (accessed March 2001).

Bradley County GIS Department (2001) GIS In Action. http://www.gis.state.tn.us/Labrary/InAction/Local/bradley_co.htm (accessed April 2001).

Brown, D. Using GIS Technology in the Development and Maintenance of a Stormwater Utility. Ohio Department of Public Utilities: Columbus, Ohio.

Data for your GIS. http://www.gis.com/data/selecting_data.html (accessed March 2001).

Data Management Support. http://www.indiana.edu/~dms/spatial/gis/prices.html (accessed March 2001).

Gillings, M.; Halls, P.; Lock, G.; Miller, P.; Phillips, G.; Ryan, N.; Wheatley, D.; Wise, A. (1998) GIS Guide to Good Practice. http://www.ads.ahds.ac.uk/project/goodguides/gis/sect53.html (accessed March 2001).

HDM. Six Steps to GIS Success. http://www.hdm.com/6steps.htm (accessed April 2001).

Imwalle, S. Maintaining A GIS: Critical Issues. http://www.awara.org/proceedings/gis32/woolprt1 (accessed April 2001).

Kartika, A.; Supriyanto, S. Distributed GIS Data Through the Web. http://www.indomap.com/geosci/ (accessed April 2001).

Korte, G. B. (2001) *The GIS Book*, 5th ed.

Laszewski, G. The Grid Information Services Working Group Participating Guide. http://www-unix.anl.gov/gridforum/gis/reports/maintenance/GIS-maintenance.html (accessed March 2001).

Mapping Strategies for Success. GIS system use and maintenance. http://www.geoinsight.com/knowledgebaase/GeoExpo99/GISSystemUse.cfm (accessed March 2001).

May, G. Using GPS to Keep Your GIS Database Up-to-Date. http://www.esri.com/library/userconf/proc00/professional/papers/PAP228.htm (accessed March 2001).

Montgomery, G. E.; Schuch, H. C. GIS Development Guide: GIS Use and Maintenance. http://www.sara.nysed.gov/pubs/gis/twelve.htm (accessed March 2001).

Natoli, J. (1997) New York State Technology Policy 97-6. http://www.oft.state.ny.us/policy/tp_976.htm (accessed July 1997).

Tax Map Conversion Guide. Maintenance. http://www.umass.edu/tei/ogia/parcelguide/Sect12.html (accessed March 2001).

Tonias, Constantine and Elias. Bridging Civil Engineering and GIS. http://www.gis.esri.com/library/userconf/proc95/to250/p250.html (accessed April 2001).

Wyoming Geographic Information Advisory Council (2000) Strategic Plan for the Wyoming Geographic Information Advisory Council. http://www.wgiac.state.wy.us/wgiac/strategicplan2000.html (accessed April 2001).

Chapter 7
Change Management

INTRODUCTION

Implementing a geographic information system (GIS) results in changes in the institutional structures and business processes of an organization. Change management refers to the administration of such changes resulting from the introduction of new computing hardware, new software development, etc., and may include a complete restructuring of the business. Hence, a framework for change management is an essential tool for any organization involved in implementing a successful GIS program.

This chapter focuses on effective methods of administering such changes. The chapter includes a brief introduction to the change management model recommended by this author and discusses change ownership, planning for change, deployment of changes, and creating a vibrant organization that can implement change quickly and effectively.

The improvisational model of change management introduced by Orlikowski and Hofman (1997) best applies to the conditions of GIS implementation. This model differs from the traditional models of change where the concept is to unfreeze, change, and refreeze the methods and processes by which an organization performs its functions. This model assumes that the work environment will be interrupted only slightly in the immediate time of implementing change. This is a valid view for introducing manufacturing changes but not for most other businesses.

The improvisational model of change management comprises three premises. The first premise is that all of the conditions and effects of the change on the specific business unit cannot be anticipated ahead of time. Second, unexpected changes will occur in related business units as a result of new technology use. For example, when a GIS is introduced to capture and maintain the location of the utility's pipelines, the information and application are now available to update the utility's financial business unit to change the way they depreciate assets. Third, implementing an information technology system will recognize new applications and new workflow efficiencies. One reason for this is that GIS technology upgrades bring in new functions that can be leveraged in the organization. Also, even if the organization does not realize new applications, the software vendors force continuous change. Accepting this model does not imply that planning and assessment are not required before the introduction or upgrade of GIS technology.

Most organizations will go through an exercise of preparing a GIS needs assessment, conceptual system and database design, and an implementation plan. Although it is true that all of the problems and consequences of the implementation of GIS cannot be fully anticipated, a good plan will identify most of them and better prepare the organization for the changes that are coming. Such plans should include a discussion of the goals of the implementation and anticipated roles of various stakeholders in making the program a success. GIS is not a technology that can be easily forced onto an organization. While some individuals will immediately see its value and adopt it eagerly, others may be resistant to the new technology and act as obstructers. A good implementation plan, created through stakeholder participation, can go a long way in reducing bureaucratic inertia.

Organizations that are highly dependent on technology to perform their business functions are well served by a management model that can easily adopt to change. Given the rate at which new technologies and software versions are introduced, a primary goal of any manager should be to create an environment where some level of change is the norm, rather than the exception. The organization must be able to quickly adopt changes that improve efficiency and quality. Organizations that can quickly adopt new technologies and the changes that come with them are better positioned in a competitive environment than those that resist change. A successful change management model enables the organization to continually improve itself through innovation.

In addition to planned and anticipated changes, other changes may unexpectedly occur either as an indirect result of the new processes or as a result of the introduced efficiencies. For example, the implementation of GIS requires data conversion. Data conversion generally consists of moving facility information from paper or digital computer-aided drafting and design files into the GIS. The most common outcome of this task is that, in the GIS, it becomes clear that the sources of data are incomplete and that information is missing. Once that is realized, the next step is typically to find out why the information is missing. After that is determined, workflows are modified to eliminate holes in the process.

Finally, managing the diverse and unrelated activities of an organization is imperative to the success of the entity, enabling it to continually improve itself through innovation. To assure continued improvement, the organization must make identifying changes a part of mid-level managers' roles and responsibilities and empower them to execute the changes. It should not be enough for the mid-level manager to deliver products and services in the same way in the same time. Mid-level managers must be challenged to innovate. Empowering execution of change requires providing the framework for mid-level managers to access technical and business process resources to bounce off ideas and to formulate specifications for technology change.

CHANGE MANAGEMENT AGENTS

The physical task of change management is different in each organization. While some organizations have offices of change management established in their human resources, information technology, or customer service departments, some rely on the internal project team implementing the new processes or technologies to take on the development and implementation of these modifications. Others bring in consulting resources to start the change process while training internal staff to institutionalize the change management framework. In all cases, a subteam of some sort must be responsible for the change management tasks. The change management team or agent is responsible for developing and deploying changes that parallel the technology implementation. The role of the change agent is to work in parallel with the technology implementers to assure that the organization does not become unduly disturbed, while at the same time assuring that the opportunity of gaining efficiency is optimized. The identified tasks are described in the following paragraphs. The key tasks for the change agent include

- Working with upper management to develop a memorandum of what the change entails,
- Establishing a means of communication,
- Developing new business processes and workflows for the organization,
- Auditing the progress of changes and soliciting feedback from the organization to assure buy-in, and
- Putting in place mechanisms to continue to leverage the technology and resulting changes in the wider organization.

The tasks identified here and described in the paragraphs that follow must happen to assure that the new technology and new business processes are implemented with minimal resistance from the organization. The negativity and resistance caused by poorly planned implementation either reduces the level of efficiencies gained or extends the time that it takes for the changes to bear fruit. It is difficult enough to train staff in new tools and processes without having the staff undermine the changes because of fears associated with the unknown.

OWNING THE CHANGE

While the idea of implementing a GIS may have risen from a group within the organization, the full hierarchy of the organization should be rallied around the technology so that the full effect of the GIS is recognized. Often, when new technology is introduced, the group requesting it owns it completely. This involves data and application development, hardware and software purchases, and staff training. A more efficient formula is to build a multi-tier project team with a number of operative units involved in decision-making and implementation processes, including business process changes.

It is important for GIS to have buy-in from the full range of upper and mid-level management. Upper management should be informed when the technology has become standard operating procedure and all efficiencies are gained through the use of GIS. Finally, the continuous involvement of management assures constant funding of the project. Mid-level management's acceptance, however, goes beyond the recognition of GIS functionality and its benefits. Because GIS applications and data are used on a daily basis for operation and staff supervision, buy-in from mid-level managers assures staff accountability in owning the changes and executing new processes. This is a long progression achieved through the involvement of mid-level managers throughout the process.

Figure 7.1 is an ideal structural hierarchy for implementing GIS where the ownership for change is distributed throughout the organization. An advisory board of mid-level managers oversees and approves the direction of the GIS implementation. The participation of an extended group of managers assures communication and creates a forum to build consensus on GIS priorities. The technical work of GIS implementation is delegated to no more than three co-project managers responsible for project issues. Advisory consulting and technical consulting resources are also essential segments in the organizational hierarchy, because the system implementation requires an infusion of veteran resources, typically from outside, for a short period of time. These resources must be fully incorporated throughout the hierarchy to assure that (a) their assistance is accepted by the staff and not seen as a threat, and (b) that they are being used and managed efficiently.

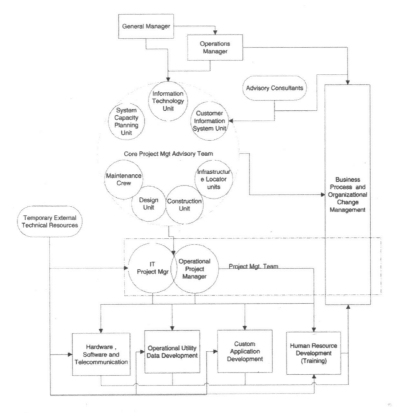

Figure 7.1   Ideal structural hierarchy for implementing GIS where ownership is distributed throughout the organization.

PLANNING FOR A SOFT LANDING

Introduction of change must be planned and done in an orderly fashion. Planning for a soft landing consists of the following four major components: (1) building a vision for the change, (2) communicating with the organization, (3) demonstrating an understanding of current business processes, and (4) building the new workflows based on a marriage of current processes and opportunities presented by the new technology.

THE VISION. The statement of vision for change is one of the first tasks of change agents. The vision should be stated in terms of the goals and view of the future. This vision must be in unison with the core ideology and the envi-

---

sioned future defined in the organization's vision statement (Collins & Porras, 1998 HBS). Specifically, the statement should be formulated to

- Be simple to understand,
- Provide a general view of the outcome of the process,
- Be usable to gauge the success, and
- Concur with the overall vision of the organization.

The statement will be used in all communication released. Staff in the organization who are eager to know about effects on their lives must not misunderstand what is to come. They will be assessing the changes occurring around them against the vision and will be making judgment calls as to how successful the implementation is. The vision can be as simple as, for example, "gaining efficiency in construction and maintenance activities in the field where crews are empowered with information". The vision for change should concur with the overall organizational vision, such as in the case of a water distribution authority where the vision of the organization is simply stated as "delivering clean water to customers in the most cost-effective and environmentally sound manner". This vision for change can be easily interpreted and adopted by mid-level managers and passed on to operational staff as business process changes in day-to-day activities. An example would include field activities and the way in which information is provided and received from the crew. It also implies a higher quality and quantity of work from the field crew. By empowering the field crew, upper management recognizes them as an essential part of the organization. This also implies that their process will be modified to quantify the productivity and quality of the field crew and accordingly structure rewards for the crew.

COMMUNICATION. The vision and change progression must be clearly and repeatedly communicated to the organization during all phases of change management. Communication is the foundation at the inception of any project and the key to its successful implementation. It should be honest and revealing. An example is a response by a company chairman in relation to a successful merger. After the merger, the chairman stated that during the process of assessing best practices of one entity versus another, their messages were brutally honest (Augustine, 1998). The communication was built on the advice of George Orwell that during the most difficult days of World War II, "The high sentiments always win in the end. The leaders who offer blood, toil, tears, and sweat always get more out of their followers than those who offer safety and a good time. When it comes to the pinch, human beings are heroic."

The first stage of communication is to prepare the employees for the changes. Employees' reactions to such changes include doubts about job security, confusion of responsibility, and future task competency. Moreover, they may feel threatened by the introduction of a new system. Some common sentiments may include

- A new tool they must learn to use, which they may feel is too complex;
- A way for the utility to reduce the number of staff, i.e., their jobs will be replaced by technology;

- An added load to their current work; and
- An opportunity to improve their job significance in the organization by making their data usable to upper management and executives.

When preparing the organization for upcoming changes, the message should be reassuring while stating the urgency for the change. State the vision, the process of GIS implementation, and the anticipated outcomes, while conveying that the change will require hard work and cooperation from staff. While initial communication will be coming from upper management, mid-level managers must be well informed about the vision for the change and the process of implementation. Middle managers will be responsible for clearing the water cooler rumors about the changes to come.

Communication for the implementation stage should include information regarding job specifications, performance, and new tasks. The introduction of GIS rarely causes job loss, yet results in higher expectations. Hence, staff members are expected to make quick and informed decisions based on infrastructure data. This expectation is typically higher for those responsible for maintaining and collecting data.

The implementation phase should have a methodology for reporting progress and receiving feedback. Progress should be reported at two levels; a high-level enterprise method for conveying the message paralleled with a method for relaying the details of the progress. At a high level, a measure of progress should be established initially, where the progress is shown as easily understood metaphors. Figure 7.2 depicts an example of an organizational implementation phase hourglass showing progress. The message may be documented in the form of an hourglass where key milestones are marked. The hourglass can be made available to the entire organization and updated when each milestone is achieved. The detailed progress reporting methodology should be established early on and remain consistent throughout the implementation.

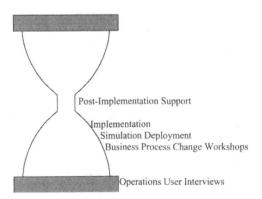

Figure 7.2   Example of organizational implementation phase hourglass showing progress.

The progress reports must be published. Detailed progress reporting should contain the following:

- Last milestone achieved;
- Success stories from past milestones;
- The next immediate milestones;
- The operative unit being affected by the next milestone;
- The process of achieving the milestone;
- A list of tasks completed to date, including the task products;
- Problems identified to date;
- Solutions to problems identified in the previous progress report; and
- Contacts managing the next milestones efforts.

It is typically difficult to establish feedback from an organization. Depending on the size, organizing an informal means of simply checking in with individuals may prove unproductive. Soliciting feedback may also be dangerous if the change management team is not empowered to solve the issues and modify the planned changes. During such circumstances, only formal venues should be created. The formality of the process will enable the change management team to raise the legitimate concerns to upper management. If the change management team has flexibility, a combination of formal and informal processes should be put in place for staff to voice their issues and concerns. People responsible for making changes must be introduced to staff and a formal contact established to document concerns. In one organization, milestones were celebrated with pizza parties where the project team reported the status of the project to all staff. It created an informal venue for reporting and hearing the concerns of the rest of the staff in the organization.

After the initial GIS implementation is complete and staff and mid-level managers have taken on the processes in their day-to-day activities, they should promote spatial data and inquire about the expansion of the application in other areas of the organization. This is achieved through showcasing internal spatial applications and applications that are deployed in other organizations. Simultaneously, the communication venues must remain active to notify the organization about upgrades and version changes. Some upgrades cause their own cycle of change communication.

Information Collection—Knowing What is There Now. It is vital for change managers to have an understanding of the organization's progress and future activities. It is also important to demonstrate this knowledge to staff. This will allow a higher level of comfort in both parties so that nothing falls through the cracks when new processes are introduced.

The process table is the basis for documenting current business processes and business requirements. The process table defines the following items:

- Task and process,
- Person responsible for the task,
- Business requirement and product achieved,

- Performance of the task, and
- Validation (duplication) of the task.

The documentation should also include

- Defining current business workflow processes targeted for the GIS implementation, i.e., how are operative units working and communicating internally and among each other?
- Modeling the data flow under the current conditions, i.e., how does information pass through the organization? Data flow through the organization is a validation step and often leads to identifying deficiencies.
- Defining the key operative units affected by GIS, i.e., who will be affected first? The success of introducing change in these operative units improves the chances of further expansion of the GIS.

This table should be developed in conjunction with the application development manual where specific GIS application requirements are defined.

The process table assures that new workflow addresses all current needs while achieving new objectives. It is important for staff to validate the table because they conduct the processes on a day-to-day basis. For example, a common problem that is missed during process analysis is emergency events. The workflow developed can begin to fail after a period of being in place, because conditions during an emergency are not incorporated. Staff from all levels of operation should review and validate the current documented workflow. This will assure that none of the operations that they perform on a day-to-day basis or during an emergency are overlooked. Several software tools enable documentation of processes and resource usage and simplify this task.

Defining the New Processes—What Changes? Implementation of a GIS requires an architectural blueprint using the business processes that the organization engages in during and after GIS implementation. It should be emphasized that this blueprint is a documentation of anticipated changes similar to the function of an architectural plan. Based on conditions and ripple effects of implementing the technology, the final processes that are put in place will probably look different from those in the original blueprint.

The blueprint is built by establishing core products and services of the business, superimposing GIS functionality, and designing the most efficient and elegant way for creating these products. The following parameters should always be applied while developing the blueprint:

- Does the new technology inherently deliver the core products and services?
- Who will use which functionality of the GIS technology to deliver the product and service?
- What additional infrastructure and tools will that person need to conduct work?

- Prioritizing the rest of the operative units based on potential long-term opportunities, i.e., which other business processes can use spatial data and applications?

All products defined in the process table must be addressed in the new workflow. Omissions should be noted with explanations.

The construction of the blueprint comes later. Basic changes are worked out on paper and during workshops to capture their ripple effects throughout the organization (Duck, 1993). It is also the time for the change team to define the implementation of GIS for a wider user community and to quantify its effects on the organization. Defining the potential for expansion early in the process prepares the organization for building a framework for a continuum of changes. This continuum is a necessity for the vitality of an organization. This topic is further discussed in later subsections of this chapter.

The best planned and documented process can still be unable to accurately address all conditions on a daily basis. The documented processes and discussions in the workshops will further be tweaked and modified during the implementation phases of the project in response to unanticipated events or missed product development processes.

DEPLOYING ANTICIPATED CHANGES

It is important to assure all policy and legal implications regarding the new processes before deploying the technology. At a utility where a new digital subdivision submission and review process was being implemented, after the first infrastructure plan was approved, the staff had to improvise and reprint the plan from their document management system so that a paper format was on file. The county's subdivision regulations contained a statement that required paper copies of the stamped plan to be on file during the construction phase of the subdivision. The regulations were later amended to drop the paper-on-file wording.

Updating the job description of staff affected by the new technology and processes is another significant task. As mentioned earlier, people are the most important aspect of a successful system implementation. Individuals affected by the change must have a clear view of what is expected of them and how they will be evaluated in the future.

The introduction of change is often the most critical part of the process. For a successful transition, the new legacy processes and systems must be operated in parallel for a length of time. This method not only validates data and applications, but also provides an opportunity to fine-tune the new workflows and business processes. Depending on the available resources and critical nature of the process, some organizations create a mock operation system completely independent of their day-to-day business, where processes are dupli-

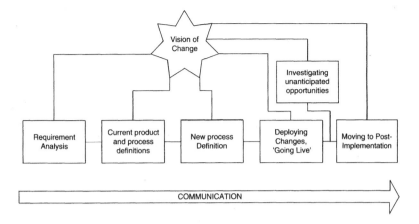

Figure 7.3 High-level schematic that demonstrates the steps for changes.

cated and products are compared between the legacy and the new processes. Others divide their daily operations between the legacy and the new processes where no duplication of work is introduced. However, it is more difficult to audit the products when processes are not duplicated.

The interconnections affecting the anticipated results of such changes occur during this parallel processing exercise. While the first reaction may involve ignoring these trends indirectly related to the technology, this silo methodology is not beneficial in the long run. Instead, the peripheral trends recognized during parallel processing should spin off their own investigation and potentially deploy new changes. Figure 7.3 is a high-level schematic that demonstrates the steps for changes. This model also recognizes the significance of peripheral trends resulting from the changes.

The successful completion of parallel processing will increase the buy-in of managers and the comfort level of staff affected by the changes. If it is not possible to swiftly resolve the issues raised during the parallel processing, a timeline should be prepared so staff may work around them before full production.

Creating a Stronger Organization

Vital organizations recognize that in the current information-based market, change is a constant. Even if business processes remain the same and no new applications are built, the information technology infrastructure is continually

---

Table 7.1 Examples of major GIS software released since 1995.

	Technology affecting GIS							
	1995	**1996**	**1997**	**1998**	**1999**	**2000**	**2001**	**2002**
ArcInfo		7.1		7.2.1	8	8.02		8.1.2
ArcView	3.2							3.3
Integraph-MGE			7			7.1.2		8
GeoMedia Pro			1	2	3	4	4 SP	5
GE Networks/	2.1	2.1	2.2	3	3.1	.1sp1&2	3.2	3.3
Smallworld	(1 &2)	(3)						
Bently-		5	5.5	5.7	7	7.1	7.2	8
Geographics								
MicroSoft	Win95			WinNT		Win2000		WinXP

changing. Hardware and software, independent of GIS, also change every year. Some of these upgrades provide opportunities for business process changes. Table 7.1 shows major GIS software released since 1995. Geographic information systems technology has changed dramatically in the past seven years. Geographic information systems vendors have experienced constant alterations every two years where the functionality, application development, or basic data model that the software used changed.

This information is significant because it shows that in addition to potential business changes, budget and expertise should be in place to navigate through such upgrades.

Frameworks put in place as part of GIS implementation should be used to build a knowledge-based organization where information flows both to and from staff. The information coming from staff is quite sensitive and should be dealt with delicately. Often, by the time ideas reach upper management, they have turned into complaints, typically regarding continuous requests for modification to things that result in no change.

Mid-level management must be empowered and encouraged to address the issues raised by day-to-day operational staff. At one organization, crew chiefs complained about spending four hours of their day opening and closing work orders in a computer system, which was taking them away from actual field-work and crew management. As time passed and the problem remained unsolved, staff complained that all technology was inefficient, adding more work to their daily tasks. They were labeled anti-technology. The problem was with the database, which did not contain the utility's assets. As a result, most of the crew spent time adding and obtaining the correct asset identification numbers for generating work orders. Simultaneously, engineers were not using the database for maintenance and design purposes. Hence, the system was not improving the efficiency of the organization as had been planned. This finding led to a business process reengineering effort, which included five operational units within the utility. Previously, the mid-level manager did not have resources on hand for addressing this issue when it was raised as an inconvenience.

To become vibrant organizations, utilities are assigning three levels of change agents. The first level includes high-level technical resources in charge of keeping up with technology and identifying candidates for the organization. Second-level change agents work closely with operational units addressing issues involving day-to-day operations affecting organizational efficiency. Technical business analysts, application analysts, and customer needs specialists are titles of resources that are assigned to operational and business units. Third are staff members who are part of the human resources division in charge of improving work environments.

This continuous attention to change internally empowers staff both at the highly technical and day-to-day operational levels. Although change is an unwelcome phenomenon, it is nevertheless constant and allows for a vibrant and innovative organization.

*R*EFERENCES

Augustine, N. R. (1998) Reshaping an Industry: Lockheed Martin's Survival. Harvard Business Review on Change; Harvard Business School Publishing: Boston, Massachusetts.

Duck, J. D. (1993) Managing Change: The Art of Balancing. Harvard Business Review on Change; Harvard Business School Publishing: Boston, Massachusetts.

Orlikowski, W. J.; Hofman, J. D. (1997) An Improvisational Model of Change Management: The Case of Groupware Technologies. *Sloan Management Review,* **38** (2), 11–21.

Chapter 8
Related Technologies and Systems Integration

RELATED TECHNOLOGIES

Geographic information systems (GIS) often connect to related information managed by other automated management systems. This chapter discusses the more common of those systems used by water and wastewater utilities and how GIS integrates with them.

Consider how employees typically access information about the utility's infrastructure. Whether responding to an incident or customer inquiry or planning work, an initial step in the process is to refer to a map. Because of this essential aspect of maps, automated systems for managing the utility's infrastructure provide some means to identify the map showing the location of a specific feature, or they have a link to GIS. Note that however powerful these automated systems may be at performing their core function, they simply cannot accomplish the essential task of almost all operations, that being getting the map.

Most of these non-GIS automated management systems are nonspatial, meaning they lack the capability to provide coordinate reference to any established spatial reference system. There simply is no way to digitally place features on a map or to view or assess features relative to other objects in the neighborhood based directly on information stored in the system's database. For example, a facility management system may store the following information describing the location of a valve: located on Main Street, two meters off the curb on the east street line of Central Avenue. This description is very precise for a service person. However, what in that location information will allow display on a map or facilitate distance measurements to nearby facilities described in other systems for managing customer information or work force scheduling? While the information is of a spatial nature, it must be enhanced with spatial coordinates to become functionally spatial within the context of this discussion and be useful in map display and system analyses. Because these automated systems lack the ability to reference a coordinate system and cannot generate maps, they cannot be used for spatially based analyses of the data, especially in relation to information in a different automated system.

Geographic information systems are the tools with the capability to accomplish the mapping and spatial analyses tasks. It is a spatial system that enables reference of items to a coordinate system, display of those features on a map, and analyses of relationships. Geographic information systems contain the base maps of the utility service area, incorporating the streets, parcels, buildings, streams, elevations, and other features typically expected on a map. Geographic information systems also contain much of the same infrastructure and customer features that are found in nonspatial systems, albeit without the level of detail found in those systems.

This chapter focuses on a utility's central GIS. This primary GIS is not the only spatial information system typically found at water and wastewater utilities. As discussed in the following sections, hydraulic models of the distribution and collection network can also be expected. These models are also spatial, having all features referenced to a coordinate system. Also, isolated ancillary GIS will likely be in use at the departmental level, each with its own database and reference base. Automated vehicle location systems also use and store positional data. These other spatial systems are discussed as related systems with the nonspatial systems.

A GIS can only display and analyze data to which it has access. Clearly, the primary GIS must work with the other automated utility management systems, both spatial and nonspatial, if the utility is to receive the full benefit of its investment in the many automated systems and information contained within their databases.

RELATED SYSTEMS. While all utilities may perform universally recognized activities when defined at a very low level, (set meter, open new account, repair hole in pipe, process "call-before-you-dig" or "one-call" ticket, flush hydrant, etc.), how utilities group those activities into departments and automated systems to manage the activities varies greatly. As discussed later in this chapter, integrating a number of smaller automated systems is complex. Vendors have responded to that complexity by bundling activities into preintegrated module suites. When doing so, they used different terminology and selected different functions to group together. One automated system vendor may include modules for work order management, asset inventory, maintenance history, and call center. Another vendor may have modules for work management systems, work force management, and facilities management. Some systems will group customer information systems and call management, while others do not. Homegrown legacy systems will have their own combination of data and activities.

The result of these inconsistencies is that utility managers must investigate the package's specific features to see which data and activities are managed when considering products from various commercial vendors. This is especially relevant if only some of the modules are being purchased, while legacy systems are to continue to manage other aspects of the overall operation. Nevertheless, the related systems described below will be in one or more of the packages. While the names may be different, the managed activity is recognizable. Functions described include

- Customer information,
- Customer contact,
- Billing,
- Automated meter reading,
- Work order management,
- Facility management,
- Asset management,
- Laboratory information management,
- Enterprise resource planning,
- Supervisory control and data acquisition,
- Hydraulic model,
- Water quality model,
- Automated vehicle location,
- Property management,
- Document management,
- Cross connection management,
- Pavement management,
- Multiple GIS, and
- Departmental databases and spreadsheets.

Customer Information Systems. Customer information systems (CIS) manage information about the services or laterals serving a property and the customers associated with the service. There may be multiple services to the property, each with separate accounts, use types, and responsible parties. In addition to information about the property being served, such as the address and type of use (residential, commercial, industrial, etc.) and the names of owners and tenants, it may include information on the lateral or service pipe (i.e., size, material, date installed), the main it is connected to, and the location of the connection, curbstop, and meter. Also included in the CIS is information on the utility's work history at the site and customer contact history. Typically, there will be feature identification (ID) numbers assigned to one or more appurtenances, such as the tap or connection, and all meters on the property. Employees using the system can access the correct record for the customer by entering the ID or the address, or by cross reference from the billing records. Information is somewhat transient, with numerous changes annually for replacement of old taps or service pipes, new installations, and changes of owner or tenant.

Primary users of the CIS are the customer service representatives as they manage accounts. New accounts need to be established when a new service is requested. All changes in the account or pipe must be entered, such as changes in property ownership or tenancy, replacement of the tap, and change in type of service. The system will keep track of all call history, work orders, and comments related to the service, customer, or property. Customer service representatives generate a large share of work orders or service requests. Therefore, if the work order processing is also automated, there likely is a link between the two systems to allow the representatives to automatically generate the work orders or service requests. The CIS is also used by field service personnel when installing or repairing service pipes, marking the facilities in response to "one-call" or "call-before-you-dig" notices, and meter setting or replacing, and by the supervisors managing those activities. Both the customer service representative and the field personnel need maps of the pertinent call or work locations. Therefore, linking GIS and CIS is a common integration.

Customer Contact. There may be a separate customer contact application to manage contact with both call- and walk-in customers. If so, information regarding contacts with the persons associated with the service will likely be in this system, rather than in the CIS.

Billing Information Systems. Billing information applications manage all details related to consumption or discharge. This includes information on the meter, meter readings, use history, billing, and payment history. Also included will be information on the property owner and the person being billed (name, address, phone number, etc.). Records are typically accessed by the name of the property owner or person billed, the address, account number, or by cross reference from the customer information records. Wastewater utilities may base the wastewater charges on metered potable water consumption data obtained from the water utility. Many wastewater utilities still use flat-rate billing and, therefore, their billing records will contain no usage data. It may contain only minimal information related to identification of the property served and the owner's billing address.

Primary users of the billing system are the employees who manage the billing and payments for use of the utility's service. Customer service representatives responding to customer inquiries about billing and payment issues and meter-related questions also use the system. Changes in the billing system must coincide with all changes in CIS regarding payment responsibility for an account. Because water and wastewater utilities often have a universal statutory lien on property, each real estate sale requires a call to the utility for current charges to be adjusted at closing. With the exception of those wastewater utilities charging a flat rate for service, the system is also used by those employees managing meters. Whether meter reading is manual or automated, meter cycles, routes, and readings must be managed. Therefore, system users include field service personnel installing meters. Because of the very close relationship of billing and CIS, these two systems are typically tightly integrated. Because wastewater billing is often based on potable water consumption, metered water consumption from the water utility may also be linked to the wastewater utility billing information system.

Automated Meter Reading Systems. Automated meter reading systems (AMR) read meters remotely and transfer the readings to a billing system. The AMR may be considered a subset of the billing system; however, it is likely to have been provided by a different vendor. The process of all AMR systems is similar. Each meter is fitted with an electronic device that will read the meter and transmit that reading. There are three general ways to get the reading back to the billing system.

- Collector units are located throughout the service area. These collector units read all the signals transmitted from meters in the local area and, in turn, transmit the collated readings back to a central location.
- A roving collector unit passes through all streets and records the transmitted readings from the units on each meter. Typically, the receiver is transported in a vehicle that drives a circuit to get within range of all meter transmitters.
- The meters are connected to the telephone circuit, and readings are transmitted to the central location via the telephone.

Automated meter reading systems are tools used by the group that reads meters and processes bills.

Work Order Management. Work order management systems (WOMs) manage the process of responding to the request for work, assembling resources, completing the work, and, finally, documenting the entire process. They vary greatly from utility to utility regarding what is included in the package and in their sophistication. There also may be a high degree of integration with other nonspatial systems, such as CIS, accounting, and payroll. Some have been integrated with GIS to the extent that they are applications of a specific GIS software product. Because many work orders deal with service pipes and meters that are either owned by the customer or are on the customer's property, a WOM integrated with CIS can be linked to the CIS data by referencing an ID, such as the tap or connection number. Of course, maintaining that link over time as taps and service pipes are changed must be managed continually.

Primary users of WOMs are persons who organize, schedule, and administer the field work force. Other users are the service personnel completing work orders with work performed, and labor, material, and equipment resources used. Additionally, those who request work would have the ability to enter a request for service, either by direct access to the work order management system or by integration with other automated systems.

For example, work orders may be requested by a customer service representative in response to a customer's call, perhaps to set a meter in a new home, by the person processing a facility mark-out job in response to a contractor's notice of excavation, or any internal supervisor or technician needing some activity performed in the field. The supervisor organizes and assigns the work, and the field personnel document work performed.

Facility Management Systems. Facility management systems are used for maintaining records of the infrastructure owned and maintained by the utility and managing periodic maintenance. Some also assist in prioritizing capital improvement projects. However, a facility management system may also have most of the functions described above for work order management, especially if WOM does not exist separately at the utility. The extent to which utility management functions are blended with functions of other systems discussed in this chapter, such as WOM, varies widely, especially with respect to management of the work. However, the primary focus is the infrastructure feature rather than any work activity that may have modified it. Where a richly attributed GIS is in place, a utility may not use a facility management system. In these cases, all facility data are often stored in the GIS and work-related activity is managed in a work order system.

For the purpose of this discussion, the scope of maintenance activity is divided into two general categories based on the type of feature involved. The first type of maintenance involves linear network features. These are the pipes, network structures (such as manholes and chambers), and appurtenances (valves, blowoffs, hydrants, etc.). They are identified by their location. The second type of maintenance involves discrete items that are commonly recognized individually (i.e., tanks, pumps, and master meters). Often, these two groups of features are managed by two separate applications provided by different vendors.

Linear networks are the infrastructure of water distribution and wastewater collection systems. Facility management systems hold in their databases the detailed characteristics of the features, such as size, type of material, lining, date installed, work history, and location. The location is typically recorded as descriptive text, identifying the structure or pipe it is connected to, the street it is on, etc. Facility management systems for wastewater utilities will commonly be associated with video inspections of pipe, so that users of the system can jump to videotape showing the respective reach of pipe. Cost data are also often included and, depending on the complexity of the system, may contain estimates, actual installation costs, and historical information.

Primary users of the linear network maintenance management system are supervisors and crew chiefs responding to pipe failure or performing maintenance work, such as exercising valves, flushing pipe, installing taps, or investigating complaints. Managers and engineers will use the facility management system to evaluate the network for prioritizing and budgeting for pipe replacement, extension, and rehabilitation projects.

The second type of facility management system deals with discrete features, such as tanks, treatment facilities, and pump stations. Items are tracked in this type of system individually, and they have their own recognized name and ID. These systems often store operation manuals, set points, periodic inspection and maintenance requirements, licenses and permits, confined space classification, emergency procedures, and a variety of other information not found in the linear network system discussed above.

Asset Management. Asset management systems are tools to maximize the cost-effective life of an asset and maintain its performance within an established range across its lifespan. Critical to achieving these objectives is determining the most effective times to maintain and refurbish components, scheduling and budgeting for that work, and actually performing it. While asset management principles have long been a factor in capital planning and budgeting for many agencies, two new drivers have focused attention on the subject, and have resulted in some new computer applications to manage the process. The U.S. Environmental Protection Agency (U.S. EPA) has proposed regulations concerning management of wastewater collection systems. The capacity, management, operations and maintenance (CMOM) rules are performance-based measures of a system's operation. Also, accounting rules for government agencies have been modified by the Government Accounting Standards Board Statement 34 (GASB 34). Both of these requirements suggest a common protocol for asset management.

Conceptually, asset management consists of a number of steps. Initially, a facilities inventory must be conducted to capture the extent of assets. The condition of all facilities must be determined and recorded, and the value of the facilities must be determined or estimated. The utility decides on a level of performance that it aspires to maintain (annual maximum breaks, unaccounted-for water, intrusion and infiltration, blocked pipe, etc.), and then determines what work must be performed and when to meet those performance standards. Based on priorities and available funds, projects are scheduled into the future.

Current sophisticated asset management systems for water and wastewater utilities have close alignment with GIS, facility management, and accounting systems.

Primary users of asset management systems will be engineers responsible for capital improvement planning. Financial managers would participate in "what-if" analyses to evaluate the effect of changing funding levels over time.

Laboratory Information Management Systems. Laboratory information management systems (LIMS) manage all information pertaining to the utility's sampling and testing of water and air. Samples are drawn from facilities such as treatment and sludge handling plants, source water intakes, and industrial customers' wastewater pretreatment discharges. They are also drawn from across the distribution and collection system for continual quality control in response to customer complaints, pipe disinfection, cleaning and lining follow-up, source water management activities, and other purposes. This sampling and review of test results ensures that water throughout the entire cycle (from raw water intake to wastewater discharge to receiving water) meets required standards for health, environment, and infrastructure protection.

Primary users of LIMS are laboratory staff. However, tests results are issued to a wide group of utility personnel, government regulators, and the general public. For example, no connections to a new water main are allowed until a lab report confirms disinfection. Therefore, personnel scheduling new taps, operations personnel who may be required to repeat the disinfection, and construction personnel who are waiting to close the excavation all need access to the information. Wastewater treatment operators need to monitor influent water quality and ensure that industrial pretreatment is maintaining standards. Water treatment plant operators need to know the raw water quality of the alternate sources and intake depths to select the sources leading to the least expensive treatment costs. Sample location is a significant item in any analyses of test results. Therefore, GIS is commonly used to relate samples to service areas, drainages, and other samples, and to evaluate areas with repeated problems. Geographic information systems are also used to graphically present results on a map for reporting to regulators and the public. A GIS is also used as a tool to manage sample sites and industrial pretreatment programs.

Supervisory Control and Data Acquisition. Supervisory control and data acquisition (SCADA) systems automate monitoring and control of the total collection and distribution system. They measure conditions and report to a central manager unit. Then, that central unit issues commands throughout the distribution and collection system to regulate the control equipment. For a typical water distribution system, the SCADA system monitors all tank levels, pumps, and pressures. Pumps are turned on in response to a drop in tank level, and alarms are reported for events such as low pump suction pressure or tank level. These SCADA systems are quite self-contained. Whether the number of points providing data is 15 or 1000, it is a relatively small number. Most of the data are historical. After all, SCADA is using only real-time (and recent past) data from those points to compute needed changes in the process to maintain it within a preset performance range. Supervisory control and data acquisition systems often copy other-than-current data to a duplicate database of the time-series data. That duplicate may be current to within a few minutes.

Typically, the sole users of SCADA are the operations personnel responsible for controlling the system day to day. While direct access to SCADA is unlikely to be provided to anyone beyond this small group (for protection of this mission-critical system), some organizations provide read-only access to the duplicate database. Otherwise, a person seeking data on system operation, either historical or current, must request a report from someone with access to SCADA.

Engineers and planners use SCADA data to validate hydraulic models. Water system modelers set up a model with SCADA data on flows, pressures, levels, and pump and control valve status at SCADA points. How well a model tracks the actual SCADA-reported conditions over time is an indication of its calibration. Total flow into or out of a service area is also extracted from SCADA for analyses of unaccounted-for water or infiltration and inflow. The key to effectively using SCADA data is to know the name SCADA uses for all of the points. Typically, access to information a week or more old is sufficient, and a greater concern for these second-tier users is ensuring that the needed data is not thinned or deleted to make disk space available for more current data.

Hydraulic Models. Hydraulic models are computer numeric simulation models of the distribution and collection networks. The network is comprised of pipes, appurtenances, and equipment. A fundamental factor of hydraulic models is the reliance on network topology (connectivity). The model must be aware of each and every connection between individual pipes. In addition to information on the pipe network, models also need flow information, which is data on water entering or leaving the network at its many nodes. Most commonly this data is compiled from metered potable water consumption from billing records or, if that source is not available, from land use and population density characteristics of the land surrounding the individual nodes. Using formulas derived from principles of conservation of energy and mass, these models predict conditions across the network based on characteristics of the pipe, structures, and equipment, and the known conditions at key points. Because of the relationship of energy to elevation and to the length and characteristic of pipe, hydraulic models must maintain data on elevation at nodes and the length of pipe between nodes. Therefore, hydraulic models have a spatial quality. They can display the network on the computer screen.

Because models and GIS both maintain substantial data (spatial and nonspatial) on the distribution and collection system, model vendors have integration tools to build models from GIS and export results to the GIS. These tools are primarily middleware (discussed in the following section). Most modeling packages extract the network from GIS, create the model, and then save the model with all ancillary files in a database separate from GIS. As an alternate to storing a separate database of the network, a modeling system may rely on the GIS database, building the model directly from the GIS data for each use session, and requiring that all necessary data be maintained in the GIS database.

Engineers have traditionally been the users of hydraulic models to perform planning and design functions. Perhaps in the future, the nature of the predominant user may shift to the operations group when full integration of GIS, SCADA, and modeling occur. While engineers will continue to use annual master models to perform planning and design functions, operations would then use the models continually to operate the system optimally and to identify overflows, breaks, and other system problems. Such an integrated model would be informed of real network conditions from GIS, real operational conditions from SCADA, and immediate customer feedback and alerts from the call contact component of CIS.

Water Quality Models. Water quality models are an extension of the hydraulic models. The hydraulic model determines flow through all model components, but does not keep track of the original source of water entering an individual pipe in the network or the way it is mixed with other water at junctions. Water quality models trace the water from the sources through the system, and keep track of all mixing and blending at junctions of pipe. They, therefore, can estimate the average age of the water at any point and the proportion that came from the various sources. As many water quality characteristics decay or develop as water passes through the system, because of time and in reaction to the type of pipe interior surface, formulas also compute these changes.

Primary users of water quality models are the team of engineers, chemists, and managers working together to ensure that water quality meets regulation-mandated limits and utility goals for both health and aesthetics. A common example would be to locate the zone of blended water from two sources with different taste and chemical composition. Fluctuating taste is noticed by residential customers, and fluctuating chemistry is a challenge for industrial customers incorporating the water to a process. The mixed zone can be moved or confined by adjusting system valves, as guided by model results.

Enterprise Resource Planning Systems. Enterprise resource planning systems (ERP) are integrated suites of applications for financial management of the utility. They may be thought of as the manager of the value chain from raw material to final product delivery. Included in the typical ERPs are modules for purchasing, inventory management, payroll, and general ledger accounting. A few large vendors dominate this software market. The common scenario relating ERP to infrastructure management concerns ordering material for a facility repair. The project manager submits a bill of material to the inventory management system. That system checks for stock on hand and, if short of material, orders, as needed, for the project or to replenish stock. The inventory management system issues a purchase requisition for the needed material via integration with the purchasing module. The purchasing module compares the material ordered to vendors, possible open orders, and approval limits. A purchase order is issued, and the vendor responds with a schedule for delivery through the purchasing module to the inventory module. Inventory responds to the project engineer with a list of available material and a delivery schedule for material ordered.

While included in this section under the heading Enterprise Resource Planning, utilities often have the components installed as separate software packages. If ERP is not used at the utility, there may be purchasing, payroll, and inventory management software packages, all from different vendors, and all trying to communicate with accounts payable and receivable, from yet another vendor.

Primary users of ERPs or separate financial management components are logically those processing data concerning the cost of resources, labor, material, and services. However, everyone with a budget or leading a project needs access to financial data to analyze the current status of his or her responsibilities, and access to the status of inventory and purchase orders.

Automated Vehicle Location Systems. Automated vehicle location systems (AVL) have merged global positioning system (GPS) technology with GIS and wireless communication to provide real-time accounting of all vehicles in a utility's fleet. A GPS receiver mounted on each vehicle receives the signal from GPS satellites and transmits its location via the selected mode (typically cellular) from the vehicle back to a base station on a preset schedule (e.g., every

5 to 10 minutes). A software application at the base station logs the position and time. The position is displayed on a GIS street map of the area. The position is related to a street address using a reverse geocoding process and presented in tables with time at locations, travel time, and computed values, such as travel speed. Work crew supervisors using the system can refer to the GIS map to see the current location of all crew vehicles.

At the day's end, supervisors can call reports detailing the time spent at all locations for any crew. The systems are very effective tools for quick response to emergency and for evaluation work force efficiency. They frequently have the added benefit of improved safety of staff working alone in isolated areas. The worker keeps an alarm button on his or her person and uses it if injured or entrapped and unable to escape. The alarm causes the vehicle to signal the home office of the emergency and the location. A common example using AVL is for the crew supervisor to respond to an emergency shutoff request by checking AVL on GIS, identifying a truck a block away and notifying that service person to perform the shutoff.

Primary users are the supervisors with crews on the road. Groups with units in their vehicles include field service workers, sample takers, material delivery trucks, fuel trucks, meter readers, inspectors, security forces (and, for a touch of equality, all supervisors with assigned vehicles).

Document Management. Document management systems keep track of file locations of electronic documents and maintain metadata (data about the data) on the documents, such as version, date created or modified, author, and keywords. The system can be quite simple; for example, it can keep track of photos and scanned images produced during tank inspections. These systems also can be extremely complex, managing computer-aided design drawings and their versions, as well as all letters and memos created within an organization. All documents of the sophisticated systems are stored on a network server accessible to everyone. When creating a document, users input its properties, including keywords, which groups or individuals are granted access, etc. Documents can then be searched and recalled based on the characteristics or properties entered. These systems facilitate sharing of documents across the organization.

A GIS often maintains links to scanned images, either as an internal GIS function or through integration with a robust document management system. The utility's facility drawings, sketches, cards, and installation records are scanned and filed. A GIS maintains a point on a base map at the appropriate location and associates it with the image filename. Clicking on the point's symbol on the map calls the image's file and causes the image to be displayed in a window. The practice is especially well suited for the large quantity of a utility's existing static documents that do not change over time, such as permits and installation records which were hand written over the years, often with a small sketch. These documents have a wealth of detail beyond which can be economically converted to digital records, and they are very valuable to field crews. The more elementary GIS functions can only go from the map point to call a related image. Document management systems linked to GIS allow searching in either system and calling associated maps or documents from the other. They are commonly used in the water and wastewater industry.

Primary users of enterprisewide document systems can be everyone in the organization who creates any letter, memo, spreadsheet, etc., and files it. With respect to infrastructure management and related activities, primary users are likely to be people regularly needing access to forms containing as-built data, permits, agreements, and filed plans.

Cross Connection Management. Found only in water utilities (not wastewater), cross connection information applications assist in managing the connections between potable and nonpotable water. These cross connections must be constructed in a way that prevents backflow of contaminated water to the distribution system. Water utilities are charged with periodically inspecting and testing these connections to ensure the protection of the distribution system. Cross connection information applications maintain information on the locations of cross connections, the owners of the connections, type of connection, inspection history, and actions taken to enforce state regulations for these connections.

Most likely, the sole users of cross connection systems are cross connections inspectors. They use the systems to manage their inspection activities to ensure compliance with utility and state regulation requirements.

Property Management. Applications referred to as property management systems may have quite different functions at different utilities, varying from management of widespread land holdings across a watershed to facility management of the utility's administrative and operational facilities. For this discussion, it refers to ownership issues of real estate holdings, the land, buildings, and easements. Many water utilities have, over time, acquired extensive land holdings for watersheds. The land was often purchased as small parcels with individual deeds, sometimes over 100 years in the past. In addition to watershed land, land has been purchased for tanks, pump stations, cross-country pipelines, and treatment facilities. Pipelines are often located within easements across the property of others. In other cases, the utility has granted or sold an easement to others for a use of the utility's property. Accurate accounting of the land is necessary for taxes, insurance, and inspection for encroachment. All of these properties, rights, and easements have value and, often, specific terms of use. Frequently, water utility land is classified by a state health agency relative to its usefulness as watershed, and the classification determines how the property may be disposed of when no longer needed by the utility. A type of a property management application can typically be found at water utilities.

Most property records include or refer to a map, often one incorporated to a legal document and on file in the public records. Linking property management systems to GIS greatly facilitates access to those maps and data. It is imperative that GIS users exercise great caution when using GIS tools to make and report land measurements related to property boundaries and area, especially when results contradict values contained in legal documents. Measurements contained in the property management system are often based on a land survey far more accurate than the GIS base map. Furthermore, they were likely established by a land surveyor with legal authority to make such determinations.

Primary users of property management systems are the personnel charged with keeping track of the property. Others seeking information include accountants inquiring about taxes, risk managers regarding insurance, and engineers regarding acceptable use of a right-of-way.

Pavement Management. Pavement management systems manage information concerning road pavements. Water and wastewater utilities typically have the bulk of their infrastructure, the pipe and manholes, located beneath paved roadways and sidewalks. Each time pavement is disrupted, whether for installation of new pipe, maintenance of old pipe, or for work on a service, that pavement is exposed to potentially life-shortening damage. Knowledge of state and local highway department plans for road reconstruction and repaving projected for upcoming years is an important factor in planning rehabilitation or replacement of the utility's infrastructure. The highway department likely has the projected paving schedules and details in an automated system. Access to the application of a different agency, such as the highway department, is a good example of integration via a Web service. A water or wastewater utility may also have its own application for managing pavement patches at excavation sites.

Primary users of pavement management systems are engineers planning major infrastructure improvements and slotting the projects in the long-term budget. Construction personnel responsible for managing pavement patches are the other primary user group.

Departmental Databases and Spreadsheets. Every department at the utility uses spreadsheets and personal databases that they have developed over time. People have databases or spreadsheets of all roofs, fences, tanks, lead paint, etc., and are using them to schedule and budget for maintenance. One person might have a database of new pipe extensions and is using it for marketing to potential customers along the route. Another might have a database of road opening permits, and someone else may have a spreadsheet listing hydrants that must be drained in the winter. Someone probably has a spreadsheet of stream flows and flood warnings. The person responsible for air permits for emergency generators likely has them in a spreadsheet. Some of the applications may be quite simple. Others were provided free by vendors of materials used in maintenance of the particular item, and include very helpful management tools.

Multiple Geographic Information Systems. There already may be multiple GIS in use at the departmental level at the utility. There certainly is the likelihood that a new enterprisewide GIS will need to receive map data from outside the agency. Because GIS is the subject of this manual, no further description is needed in this section on related technology. Issues related to integrating multiple GIS are the subject of the following sections.

Primary users of existing smaller GIS could be everyone who spatially analyzes those features within the scope of their responsibility. For example, a forester managing a utility's watershed may use aerial photography of a land holding obtained from some other agency and an inexpensive or free GIS to develop quantities and a bid specification for a timber sale.

GEOGRAPHIC INFORMATION SYSTEMS INTEGRATION BENEFITS, PROBLEMS, AND OPPORTUNITIES

GEOGRAPHIC INFORMATION SYSTEMS INTEGRATION BENEFITS. Integration is simply the sharing of data and functions among various systems. Sharing with GIS provides opportunities to extend the value of any other system's data, such as enabling

- Rapid access and display,
- Spatial analyses,
- Elimination of duplicate data entry, and
- Improved database administration.

Rapid Access and Display. The GIS integration benefit most apparent is the ability to rapidly access and display up-to-date information from all associated automated systems on a map. As noted in the opening paragraph of this chapter, the initial step of most utility work activities is to obtain the map. A GIS is a common portal to all maps, including maps generated from information within the user's primary automated system or some other system. When integrated, viewing another system's data requires neither logging on nor being familiar with that system's user interface. All maps are viewed using the common GIS tools. Many utility workers spend much of their time answering questions from customers, contractors, engineers, attorneys, managers, board members, or city council members. Integrating the many systems results in faster, more accurate answers. Examples of display and query functions supported by accessing another system's data include

- Customer information. An engineer planning a preblast survey uses GIS to view and list owners of all properties near the construction site.
- Laboratory information management data. A construction crew ready to backfill a disinfected new pipe and a customer service representative responding to a request for a new service both use GIS to locate the pipe and check LIMS for adequate chlorine residual and authorization to put pipe in service.

- Hydraulic model information. When reviewing a prospective new customer's request for service, the reviewer accesses the new record in the CIS and then locates the homeowner's property in GIS. The reviewer checks a layer of pressure contours, exported from the hydraulic model, to ensure that an acceptable level of service is possible.

Spatial Analysis. Automated systems have powerful data query and analysis capabilities. However, those capabilities are extremely limited when dealing with questions of location, and they have no ability to relate spatially to items not in that system's database. A GIS provides the capability to analyze the data spatially. The spatial analyses may be of data within the user's primary system or may be used to compare data of multiple systems integrated with GIS. Examples of spatial analysis include

- Laboratory information and facilities management. A team investigating the benefit and cost of cement-lining mains uses GIS to explore any trend of customer complaints and high water pH (from LIMS) near recent cement-lining areas (from facility management) through subsequent years.
- Facilities management. A person preparing annual reports queries GIS for summaries of pipe maintained in the facilities management system, compiled by administrative areas defined in GIS.

Duplicate Entry Avoidance. A price is paid for maintaining duplicate data in separate systems as an alternative to integration. Clearly, there is double the cost to enter new data and updates twice, and with each duplicated data entry, the potential for errors and inconsistencies multiplies. With GIS integration, duplicate entry is reduced. Examples include

- Billing data. Engineering uses consumption data from the billing system for loading hydraulic models.
- Facility data. Facility management system modifications directly update the GIS.
- GIS data. The GIS distribution and collection system networks are exported to the hydraulic model.

Not as apparent as the cost of double data entry is the problem of the data being different. Discrepancies will occur when updates are performed at different intervals and as mistakes are made in one system or the other. When discrepancies surface, work stops for a review of which system has the correct and complete data. Either the issue is investigated, likely by field inspection, or one system is declared more reliable than the other. Perhaps more troublesome for the managers of the systems is the example of facilities management and GIS both compiling totals of infrastructure in place and coming up with substantially different totals. Because taxes and various municipal charges are based on these totals, it is critical to know which one is to become the official one, and who will make that judgment call. By eliminating the need for duplicate data entry to multiple systems, GIS can greatly reduce the cost, duplicated effort, and the potential for errors and inconsistencies.

Database Administration. Integration of smaller local databases with larger enterprise systems often results in better management of the database. The local database owner likely has other higher priority tasks than administering security and backup. A system administrator within the information technology department manages these activities as a core responsibility.

POTENTIAL GEOGRAPHIC INFORMATION SYSTEMS INTEGRATION PROBLEMS. Along with the benefits of GIS integration come some problems that must be addressed to ensure the continued viability of the utility's data. While problems related to different coordinate systems of spatial datasets will be apparent at the first attempt to display together, other problems are insidious, and may not be realized until a valuable database is irreversibly damaged. The following pages describe some of these pitfalls and ways to avoid them.

Coordinate Systems. When referring to spatial data, it is common to think location and assume it is quite straightforward. Unfortunately, a location has many factors that must be considered during integration of different spatial datasets, one of which is the coordinate system. It is likely that GIS will be called on to integrate data spatially referenced by latitude and longitude with data using State Plane or Universal Transverse Mercator (UTM) (two other common coordinate systems). As long as both coordinate systems are known and are industry standards, a modern GIS will be able manipulate one or the other so they will display together. If not, maps referenced to different coordinate systems simply will not be able to be integrated.

Format. Information in a database conforms to whatever format was selected by the developer. Suppose the existing work order system was developed in-house some time ago, and the programmer made distance a text field so that it would accept almost whatever was typed in, and no limit on the measurement was established. Now distance is recorded in many ways, including "3 m", "12-meters", and "fortie-five". Further assume that GIS has distance as a numeric value. Before any distance data from the work order system can be used to update GIS, it must be converted from text to numeric, and all variations must be converted to the conforming format, like "30.00". Integration involving incompatible data formats must provide a format conversion step to occur each time data are passed. The misspelled number (fortie-five) must be scrubbed (cleaned of errors) in the work order system so that the automated conversion routine will recognize it.

Definition and Domain. Integration of datasets requires that every definition be consistent through the integrated systems. For example, a pipe will go through many phases as it moves through proposal, design, construction, and active use, to replacement or abandonment and possibly finally removal. All

---

users of facility management, GIS, work order, and LIMS must agree on precisely when a pipe changes status, and during which phases each feature will be included in the individual systems. With respect to changing status from active to abandoned, does the change occur when an isolating valve is shut? Or, is a physical separation (pipe removal) required? If a pipe is removed or replaced, does it remain in the system, and is status removed or replaced? There will be many similar cases of different criteria in different departments.

One system may accept the terms easement and right of way interchangeably, while another may use both, but distinguish between them. The operations and engineering departments of a water utility may have different definitions of service area hydraulic gradient, and will maintain different elevations in their separate systems. One group will consider service area to be a pressure zone while another considers it to include everything downstream of a master meter. If wastewater is managed in the same utility, that group will have a different use for the term. Most systems will maintain domains of acceptable values for data items for quality control purposes. Just as with feature definitions, the domain lists of allowable values must also be synchronized. If facilities management updates the domain of manufacturers for hydrants based on a change of suppliers, and does not inform GIS, a newly added hydrant will not be accepted by GIS during a periodic update. The list of these type anomalies is long. Integrating these systems requires that they all be resolved, which then paves the way to realize the many benefits of GIS integration.

Precision. Precision is discussed in the Data section of Chapter 5 under Spatial Accuracy. Within the context of that section, precision of spatial measurements refers to the range of numeric values that can be stored in a coordinate (double precision coordinates use eight bytes of data for each dimension.) As used in this chapter, precision has a broader context to also include the fineness of an original measurement (i.e., was the point's coordinates originally measured to the nearest or one meter, etc.). Mingling data of mismatched precision into a single set has potential for significant damage to the more precise dataset that likely was very expensive to create. The ability of GIS to zoom in and out of a map disguises the scale at which a map was created and its original intended use. For example, an operations group may have a map of the entire town created for use and display at 1:10 000 scale that is used to show generally where major facilities are located. This map may show a lift station at 30 meters from its true position, but it is perfectly well suited for its intended use. However, GIS allows a user to zoom in and display this map at a much larger scale. If this map data were merged together with data of another map of the collection system, one created at 1:500 scale and accurate within 1 meter of true position, the combined data would be incoherent. It may well place the lift station across the street on top of someone's patio. This example is extreme and therefore easy to avoid. More insidious would be to accept a file containing updated information on manholes based on a field survey, and to overwrite an existing file with the new, including new locations. If the old locations were from photogrammetry captured at a scale of 1:1000 and accurate to 1 meter, and the file had new locations for some manholes, captured by handheld GPS (i.e., plus or minus 5 meters), the free file was no bargain. Once data sets are merged, a user has no way of knowing what the level of precision is for the location of any item being investigated, and all items fall to the coarsest level of locational reliability.

Another precision issue is the inability to merge two distinct groups of data and users within the utility because of very different map bases. A utility possibly has a land management group using small scale maps (small because ratio of map distance to actual distance is a small fraction, i.e., 1 m = 20 000 m is a ratio of 1/20 000, a small number), possibly with aerial photography, all referenced to the U.S. Geological Survey (USGS) quadrangle sheets. These land management GIS maps will integrate well with much environmental data and maps created by other local and state agencies dealing with land use, planning, and environmental issues, because these agencies also use the USGS maps as a base. However, they cannot be integrated with the detail maps of the distribution and collection system (created at a scale of 1:1000) to display the detail map feature against the aerial photo background at large scale. The aerial photography simply will not line up with the streets and buildings of the detail maps.

Precision is also an issue for GIS expected to be used for building hydraulic models. Models depend on a pipe network's topology, which is the relationship of features to other connected or adjacent features. While models will maintain topology in a table of all pipes, listing the other pipes or features they are connected to, GIS, with few exceptions, maintain topology spatially. In other words, if endpoints of multiple pipes occur in the same space they are assumed to be connected. Therefore, when extracting a pipe network, most models must build topology based in that spatial coincidence. One of the issues encountered building models from a GIS concerns how precisely the GIS was digitized. If the pipe network was digitized at a smaller scale, and pipe looked continuous on GIS maps even though the endpoints have coordinates some distance apart, the model may assume they are not connected. If, to properly connect these pipes, the snap distance (a user-defined parameter that pulls points in close proximity together) is increased sufficiently, for example, to 2 meters, all pipes with endpoints within a 2-meter proximity of each other will be connected. This will likely incorrectly link pipes that in reality are not at all connected even though they are within 2 meters. Thus, the model will not accurately reflect actual conditions. If an existing hydraulic model was originally built independently from the GIS, it likely has very carefully constructed topology stored in a model table. This topology can be used to guide repair of an imprecise GIS pipe network, providing the platform for accurate future extractions of the network.

Reliability. Before making decisions and taking actions on presented data, the reliability of the data must be known. How likely is it to be accurate? If an automated system has great functionality but the data are often simply missing or, worse yet, wrong, the utility supervisors and workers will quit using the system. The GIS will sit unused as they pull out marked-up old paper maps and files. Reliability can be reinforced by ensuring that any data imported to GIS from another system meets the accuracy standards of the GIS, and that adequate controls exist in the other system regarding access and permission to input data. When importing bulk data from a foreign system outside the utility, the GIS manager must be vigilant to insure that the data meet the system's standards, and must keep track of the metadata (the data about the data).

In addition to data reliability, there is an issue to consider concerning hardware and network reliability. Depending on many variables, transferring a stable local application to an integrated environment on the network means a disruption of the network or that system with the needed data may render the data unavailable.

Avoidance and Management of Integration Problems. Listing the above potential problems in no way suggests any of them are likely to interfere with a utility's integration project. However, awareness of potential problems leads to timely discovery and hopefully avoidance. If unavoidable, identifying the problems leads to assessment, planning, budgeting, and resolution. By addressing these issues with vendors or employees proposing integration of automated systems, utility managers will be helping to ensure that their resulting project is properly scoped and remains within the proposed schedule and budget.

DATA SHARING OPPORTUNITIES. Enthused by the benefits of integration and forewarned of pitfalls, where does one look for opportunities? Integration opportunities will exist wherever the same data are maintained in multiple datasets, wherever a user of one system picks up a map to locate information managed in a different system, and where the value of the benefits exceeds the cost and risk of implementing the integration.

Data Common to Multiple Systems. It is easy to find examples of data that are used by multiple departments of a water or wastewater utility. An example is the diameter of a water service or a sewer lateral. Field service groups need the diameter for maintenance activities and maintain it in the facilities management system. Customer service needs diameter when discussing the service with the customer and maintains it in the CIS. A GIS maintains service diameter as a frequently used attribute of GIS features. Billing will likely base a portion of the fee on diameter. A records group compiles distribution statistics on services by diameter for annual reporting. All of these users may have a file or database that includes information on service pipe diameter.

Data Present Only in a Single System. Some information does not have problems of multiple definitions because it exists only in one system. A good example is water consumption data. Details of consumption can only be found in the billing system. Using consumption data in any engineering analyses of the distribution system requires integration if manual data reentry is to be avoided.

Land Base Sharing. Perhaps the most significant effort at sharing of spatial data involves base map features and topography, planimetric, and cadastral data. It is very common for the base maps to be managed externally to the water or wastewater utility, by a parallel governmental agency of the city, county, or region. When land base is maintained by an umbrella organization and shared by a community of users, costs are obviously significantly reduced. Disputes between agencies can become heated during resolution of a common land-base precision standard and discussions on features to capture. Typically, water and wastewater utilities need a greater precision database than planning, public safety, and other agencies with lesser need for accurate positioning of map features.

FINDING INTEGRATION OPPORTUNITIES. An important question to answer at the start when considering GIS integration is whether one is interested in simply integrating existing systems at a utility with GIS or whether one's interest is broader and includes all spatial data sharing potential. This section will assume the latter and look at all relationships between activities and data. Existing related systems may already have associations between each other. Concluding that a change in one of those systems is needed to facilitate GIS integration may negatively affect its other associations. Although not required, GIS integration is often best considered within an overall information systems master plan. An updated master plan will likely incorporate modern business process concepts and business workflow to address communication of information between workgroups and individuals. Just as the flow of water through a utility's distribution system can be modeled to understand the current process and to develop and implement changes (refer to the Hydraulic Models section of this chapter), so too the flow of work and data through an organization can be modeled to achieve understanding and to develop a new enterprise architecture for the business process. Chapter 5 discusses the software design development cycle, the software modeling languages and components of application development. Chapter 7 discusses documenting current business workflow processes and managing the inevitable changes. Both chapters apply to finding and implementing system integration.

Successful GIS integration projects are accomplished by teams with the requisite expertise in systems integration, software development methodology, commercially available utility applications, and utility operations. The many aspects of integration cannot be covered in one section of a chapter of this manual. Nevertheless, there is value in discussing a high-level approach to assessing integration opportunities. Utility managers possessing an understanding of the overall process have the resources to select appropriate consultants and to participate effectively as the project progresses. A process for initially identifying integration opportunities is described below in six steps.

(1) Identify work activity,
(2) Identify data,

---

(3) Develop matrix,

(4) Filter matrix for GIS activities,

(5) Assign communication type, and

(6) Determine use frequency.

Supported by knowledge of GIS integration opportunities at the utility and with a summary view of the integration requirements and options discussed in the following section, a manager is ready for meaningful discussion with vendors of GIS integrated suites and GIS integrators.

Identify Work Activities. If a group of managers with similar roles at different utilities were asked to provide an organizational chart down to the departmental level, and then list the functions of each department, it would likely be that the organizations would be very different, and that the work is apportioned among the groups quite differently. One water utility organization may have personnel that set meters in a customer service group who are tightly integrated with personnel managing customer accounts and billing. Their work activities may be included in a CIS that manages work orders for these individuals. Another utility may have these persons as part of an operations maintenance group, working closely with crews operating valves, and maintaining equipment. Their work orders may be issued from a work management system separate from customer information. While a wastewater utility would not deal with meters, they would have tasks assigned to groups to deal with all aspects of managing backups, blockages, sags, or cracked pipes. The point is that all organizations will universally have essentially identical activities at the most elemental level for water or wastewater. After all, they are achieving the same larger function, which is providing a water or wastewater service to the community. Therefore, all activities should be listed at a quite fine level of detail. If listed in a hierarchy, they can be rolled up or expanded as needed during analyses. For example, a water utility may have a distribution operations department that can be broken into process control, scheduled maintenance, and event response. The activities in event response might be

Distribution Operations
    Event Response
        One-Call Markout
        Leak Investigation
        Valve Operation

Identify Data. What data are captured, created, and maintained by the utility? There are many categories of data. What are the things that make up the company? These things are tangible, such as pipes, buildings, equipment, customers, and material in inventory. There are also intangible things, such as billing accounts, flow data, general ledger, and work histories. As with the activities above, providing a hierarchy allows the detail to be rolled up for higher-level review. For example, the initial list for pipe and other network items may be grouped as follows:

Facilities
  <u>Linear network</u>
    Pipe: Type, material, class, size, lining, date installed, joint type, proj #
    Valve: Type, class, material, style, size, direction, manufacturer, # turns
    Manholes: Type, size, barrel, material, depth, date installed, project #
    Appurtenances: Type, size, material, date installed, project #

A thorough assessment of the data would examine each of the items and identify what systems they are stored in and their format, precision, rules, associations, etc. However, much is also gained by the higher-level initial review.

Develop a Matrix. What data does an activity use? Integration potential and problems obviously only exist in instances where the activity uses the data. A matrix of the data and the activities that use it identifies activity and data associations to review further. Figure 8.1 shows such a matrix. For example, discussing a customer's question or problem over the phone might require access to many types of data. The call may involve a water meter, a sewer overflow in the street, a change of name, or a billing problem. However, managing a call will not likely require access to SCADA data. Therefore, there is no association of call management to SCADA data. Completing this task will identify all instances of activities using common data.

Filter Matrix for Geographic Information Systems Activities. Although useful organizationwide, the developed matrix includes associations that are not remotely related to GIS. As this process concerns integration with GIS, the next step is to identify the associations where a map is used by the activity to find the data. The matrix should be filtered for only those associations that are likely to benefit from and be involved with GIS integration. Some data lookup does not involve a map. For example, a supervisor of a pipe break repair job needs to know availability of parts. The supervisor would use the inventory management system (or ERP) to check availability. A GIS would be of no benefit. However, suppose the supervisor needs to know details of any critical customers in the vicinity of the break. This need requires a map to show the location of nearby critical customers and access to customer information for details on their special needs. This indicates a probable use for integration of WOM and customer information that would benefit from GIS integration. This process results in a matrix of activity and data associations that are facilitated using maps, and they are the only ones to be analyzed further. The matrix value could be enhanced by coding users and the type of usage within the cells. This would indicate which groups are using the data, and whether they need ability to create, read, or delete the data.

Assign Communication Type. How immediate is the need for accurate data? As is revealed in the Integration Options section, the complexity of an

DATA	Activities: Opns & Maint			Customer Mgmt			Planning & Engineering			Water Quality Management					Administration & Finance					
	Process Control	Scheduled Maintenance	Event Response	Call/Query Mgmt	Customer Info Admin	Billing	Asset Mgmt	Unaccounted Water Analysis	Hydraulic Modeling	Sampling & Testing	Source Protection	Water Quality Modeling	Cross Connection	Land Management	Purchasing	Inventory	Property mgmt	Document Mgmt	Finance	Risk Mgmt
BASE MAP DATA																				
Topographic																				
Planimetric																				
Cadastral																				
Remote Imagery																				
FACILITY DATA																				
Linear Networks																				
Plants																				
CUSTOMER DATA																				
Customer Information																				
Metering & Billing																				
Contact History																				
OPERATIONAL DATA																				
CBYD / One-Call																				
Scheduled Maintenance																				
SCADA																				

Figure 8.1 Example of a user matrix.

integration is greatly influenced by the type of communication required. A communication that can do well reading information that is a little stale is much easier to manage than one that must have the most up-to-date information from another system. Will one system periodically synchronize with and update data in another system? That is less difficult than a communication that must make a live update to data in another system.

Determine Use Frequency. What is the frequency at which this association is used? Know where automation and integration will provide the greatest benefit. Be sure to include in the integration project those communication associations that are repeatedly used, have high volume, and immediacy. Be ready to abandon a low-priority association that is seldom used, and that increases the integration complexity disproportionately.

The outcome of this process is a general understanding of GIS integration opportunities at the utility. A product of the effort would be the matrix revealing the data and activities associations with spatial characteristics at that particular organization. Perhaps one association would be at a higher level, with both activities and data items rolled up to major categories. Another association may be subsets of associations with substantial connections exploded to the finest detail, if time allowed matrix development to that extent. Results of the analyses of types of communication and frequency of use can be coded on the matrix cells to show the combined result in a single view.

INTEGRATION REQUIREMENTS AND OPTIONS

REQUIREMENTS FOR INTEGRATION. Integration of GIS and related technologies can take a variety of shapes and forms. However, there are a few essential elements common to all integrations.

Communication. Some means of communication must exist. Whether exporting a file to a floppy disk and carrying it to another computer to import, automatically passing messages back and forth between systems, or reaching into a separate application's database, the systems must communicate.

Compatible Format and Range. As noted in the section discussing integration problems, data within an integrated system must be compatible in format, range, definition, units, and precision. Some problems will require data scrubbing (cleaning of errors) before integration. Some inconsistencies may be allowed to remain, as long as the data are converted during each transfer between systems. Chapter 4 discusses data integration issues, identifying common data relationships that should be taken into account because they affect integration opportunities.

Synchronization. Automated systems typically are fairly dynamic. Simply establishing a link between a pipe record in the facility management system and its counterpart pipe record in GIS is not sufficient. The systems must be kept synchronized. Both GIS and facility management must keep track of and accommodate changes made in the other system. If a main break results in a pipe being removed and replaced, all systems maintaining information on the pipe must be updated. For example, break repairs are entered into the facility management system, resulting in deletion of the removed pipe's record and addition of a new record for the replacement. A GIS must be notified to change its records accordingly.

Referential Integrity. Referential integrity is the characteristic of a system describing the extent and accuracy of associations among related features. For instance, a cleanout does not exist in a real sewer system without an associated pipe, nor should a meter be located on a service without an associated billing account. These real-world associations should also be maintained in the automated information systems. To perform their functions, all of the systems maintain an extensive group of tables that keep track of associations of individual data records of the other tables. Modifying a database item mandates that these tables be updated concomitantly to ensure that internal referential integrity is maintained.

For example, some collection system GIS include manholes as a feature with number of connections as an attribute. If the utility cuts and plugs an

---

abandoned pipe at the manhole and the facility management system notifies GIS of the work, deleting the abandoned pipe record in GIS is not sufficient to maintain integrity of the association. A GIS must also modify the manhole record and the table keeping track of the relationship between pipe and manholes. Rules of connectivity and relationships must be developed and followed automatically to ensure continued integrity of the total database.

Creating the needed connectivity and relationship rules is a task for the developer programming the interface. Applying the rules is an automated activity of the applications. Therefore, the burden of creating and maintaining these rules over time is substantially on the developer. Checking all relationships and applying them becomes a significant burden on the computer processor as the size of the databases grows. The burdens on developers and on processing time are drivers in deciding what type integration is best for a particular utility.

Design. All mechanisms of communication, data conversion to compatible formats, synchronization, and definition and maintenance of referential integrity must be created by a software developer. The work involved to develop, test, and implement the integration can be substantial in terms of time required, cost, and risk. All of these factors are reduced by using reusable configurable modules in lieu of custom integration of existing applications. Utility management software vendors have bundled related applications into modules of management suites, and have preintegrated the modules. Utilities select which modules to include, and the vendor (or an approved consultant) configures the modules to the particular client's organization. The total burden of custom integrating legacy systems will often far exceed that of using a modular preintegrated package, including purchasing, configuration, and data migration.

INTEGRATION OPTIONS. This chapter defines the term integration very broadly to mean all forms of communication between separate automated systems in use at a water or wastewater utility. In the context of this definition, integration options discussed in this section cover the range from the simple passing of a floppy disk between two computers to the complex level found only at very few large utilities today.

Periodic Data Update. The simplest means to integrate systems is to periodically export a data file from one system into the other system to update its database. As an alternative, an intermediary database is maintained on the network, and an automated, periodic updating routine is programmed. For example, assume that the CIS is a legacy automated system that is updated with changes made to a customer's water service pipe, and that GIS also maintains data on tap ID, installation date, size, and material. A procedure is programmed in the legacy CIS to periodically export a file containing all modifications since the last export. The program selects those records since the last operation and, if necessary, converts formats for compatibility and exports a file.

However, not all data in the new file will find its related record in the GIS. The issue is synchronization. The GIS operator also needs a list of all additions and deletions from the CIS. Using that list, the operator adds all new taps and services to GIS, linking the feature to the associated record of the imported data. The operator also goes to each tap in GIS that was destroyed and changes its status, adds the new tap, and associates the old service, the new tap, and the imported data record.

- Advantages. This method does not require extensive programming within the old legacy CIS that may no longer be well documented. It also does not require detailed rules for updating the GIS.
- Disadvantages. Any changes made by the GIS operator in target GIS fields will be overwritten at next periodic data import.

Customized Integration. Integrating formerly stand-alone systems can be a daunting challenge. Each integrated feature of one system must be mapped to the feature in another that contains data it seeks or to an interface of the other system. As the number of systems being integrated increases, the number of interfaces and procedures to define and manage also increases, and these increases can be exponential. The customer representative takes a call regarding a customer's service and automatically opens a record in the call management system. The representative asks for an account number and street address and, using that information, accesses both the CIS and the billing system. Through integration of CIS with GIS, a map of the customer's premise is displayed on the representative's screen. Using all accessed data, the representative can efficiently address the caller's issue. Figure 8.2 shows an example of a customized integration system.

Each of the communications between systems required a programmed procedure to be accomplished. If the target system had an open database, the procedure could directly access the proper records. If the target system maintains the data in proprietary files, then the procedure would be a message requesting that the other system obtain and report the data of interest. Because one system asks for data from the other by the name of its field, changing the name or location of data in one system requires corresponding adjustments in all other systems with coded reference to it.

The foregoing example required integration allowing a system to only read data from other systems. The challenge is in keeping the systems synchronized so that the related records remain associated. A bigger challenge is to allow one system to update data in the other. If duplication of data entry and update effort is to be eliminated, new data and modifications entered into one system must have the capability to flow through to the others. Effecting changes in related systems requires expert knowledge of both systems to ensure that referential integrity is maintained during an update invoked by a foreign system.

- Advantages. Customization may be the only realistic option for integrating legacy systems. If both systems are well known by the utility's information technology department, internal programmers can develop the

---

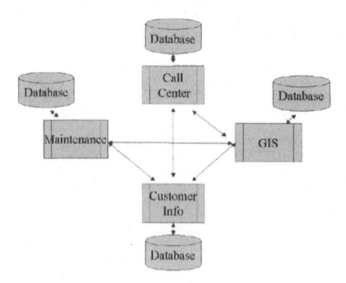

Figure 8.2   Example of a customized integration system.

integration through consultation with nearby users and without the need to learn the organization.
- Disadvantages. The time and expense required, both to implement and to maintain the integrating application over time, will typically exceed the cost of converting from a legacy system to configurable, preintegrated modules.

Middleware. Middleware is software used to process data from one system to be useful in another system. Use of middleware is common where data needed by one system are stored and updated in another, but the required data needs substantial reconfiguration before it can be used in the requesting system. The most common example is the use of GIS data to build and update the pipe network of a hydraulic model. For a water distribution network, the pipe, valves, pump stations, pressure reducing valves, tanks, and other network features are all stored and kept up-to-date in the GIS. Unless the model is also updated separately and in parallel with GIS as changes are made to the distribution system, these GIS updates must be transferred into the model. However, because the functions of GIS involve spatial analyses and display, and the function of the model is numerical analyses of a network, the data stored in GIS does not include all of the information that is needed for the model. A GIS will not likely include the pipe's friction factor in stored data. In the water network example, control valves will not have data on set points, direction, or flow coefficients. Tank records will not include shape-to-volume relationships, and pump records will not store pump curves. While some GIS maintain network topology, most currently in use do not. All of these categories of information are needed by the model.

Middleware is an application, typically developed within the user application (in the above example, within the modeling application) in close cooperation with the GIS vendor. Continuing the hydraulic model example, typical middleware would extract a GIS dataset and compare it to a dataset from the hydraulic model that is to be updated; typically the middleware is capable of importing and exporting GIS data in the GIS proprietary file format. Additions, deletions, and changes influencing flow characteristics, such as relining, would be processed. Pipe reaches with similar flow characteristics are consolidated. Data not stored in the GIS, but needed in the model, are stored in the middleware's file and added and adjusted in response to input from GIS. Finally, the process exports a model file to the hydraulic model application. The middleware can also export model results back to the GIS for display. Another example of middleware needed for model building processes consumption data from the billing system to normalize reading cycles and dates to a common time period. Middleware performing this task would reside between the billing system and GIS. Figure 8.3 shows an example of a middleware system.

- Advantages. By organizing and processing data from the source database before transferring to the target, middleware allows the source to remain somewhat nonconforming to the needs of the target.
- Disadvantages. The application must perform the data processing functions each time data are transferred between systems.

Common Spatial Database. Until recently, GIS spatial data could not be stored in the same open database table as attribute data. Spatial databases now allow the spatial component to be a feature property and to be stored as a field of the feature record in an open accessible database. Therefore, all information related to a utility's infrastructure can conceivably be stored in a common database, and accessed to be read or modified by any of the automated systems. However, this concept puts the rules of connectivity, relationships, and permissions at the database level. These rules become increasingly complex as multiple systems have permission to modify the database. Version management of long transactions extending over the long term (e.g., the life of an engineering and construction project) further complicates the data

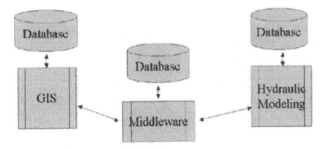

Figure 8.3   Example of a middleware system.

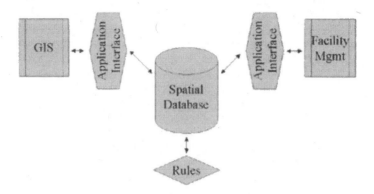

Figure 8.4   Example of a common spatial database system.

management. This option for integration is relatively new. Figure 8.4 shows an example of a common spatial database system.

- Advantages. Data duplication is minimized. All applications have direct access without use of another system's interface.
- Disadvantages. Relationship rules and connectivity become complex. Decreased GIS performance may be noticed because of the additional complexity of the data source.

Preintegrated Geographic Information Systems Extension. Given all of the complications of integrating existing systems from different vendors, one solution is to bring the related applications under the umbrella structure of the GIS. A GIS is the intuitive choice because it is so universally involved. Almost all infrastructure data have a location component. With this option, all data are maintained within the preintegrated databases and managed using procedures developed using that GIS's program language. The developer of the modules is typically a business partner of the GIS vendor. This is a common approach to integration in the water and wastewater industry. For example, integration of the call center, work order, and facility management systems with GIS allows the call taker to display a map of the caller's premise and issue a work order (e.g., "check for missing manhole cover") while on the phone. Work details and details of changes to the infrastructure (e.g., "broken manhole cover replaced") are recorded with links between the different module tables. Because the modules were developed by a single vendor having a business relationship with the GIS vendor, the integration issues were addressed as the modules were developed. The modules and the umbrella extension are configured by the vendor or a consultant to fit the utility installing the system. Figure 8.5 shows an example of a preintegrated GIS extension system.

- Advantages. Data duplication is eliminated. All applications were developed as coordinated effort. Implementation is less risky because of preintegration.

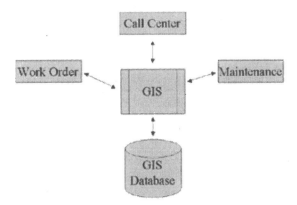

Figure 8.5   Example of a preintegrated GIS extension system.

- Disadvantages. The quality and level of support for the entire package are subject to the general health and viability of the underlying GIS vendor.

Preintegrated Suites and Connectors. Just as modules can be developed within the GIS environment, they also can be developed in a preintegrated suite separate from GIS. Common examples of this option are the ERP and facility management suites. These suites have modules tightly integrated to use a common database, ensure referential integrity, and facilitate implementation and administration. Configuration, rather than customization, is the guiding principle. As modules are purchased and brought into the utility's system, they are configured to accommodate the utility's individual situation. Given the complexity of the total suite and the risk of data corruption should its integrity be disrupted, suite vendors likely will prohibit direct access to its database. Instead, they provide connectors or portals, through which other applications can communicate with the suite. To facilitate integration with GIS, major suite and GIS vendors have coordinated to develop these connectors between their software environments. Figure 8.6 shows an example of a preintegrated suites and connectors system.

- Advantages. Risk is minimized by using preintegrated suites. Interfaces with GIS and other suites are planned to be configured for the individual utility.
- Disadvantages. Suites may include more than needed or desired by the utility.

Publish and Subscribe. Going another step beyond the preintegrated suites with connectors to communicate, an integration bus can be added to facilitate multiple suites and systems. Note in the schematic in Figure 8.6 that the two suites from the preintegrated suites and connectors example above are joined by a legacy CIS and the GIS. For large organizations, use of an integration bus becomes an option. The bus is the product of an enterprise application inte-

---

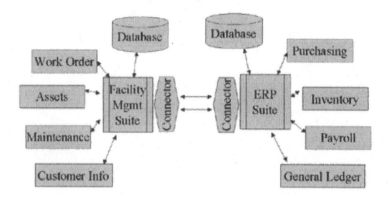

Figure 8.6   Example of a preintegrated suites and connectors system.

grator vendor. Components include portals through which individual products or suites of modules communicate (similar to the connectors of the preintegrated suites option) and a message broker module to monitor and direct message traffic.

During a common incident (e.g., a contractor excavating in the street rips out a section of sewer, including a lateral), all systems shown below will be affected. Pipes involved will be initially identified in GIS, the customer involved will be identified in CIS, and materials needed for repair will be identified and located using the inventory module of ERP. When an event occurs in any of the systems (assuming, for this example, that the facility management system has an application for pipe repair that requests availability of needed parts), a message is published to the integration bus. Other systems with interest in this type of message have subscribed to be notified of the event. In the present example, ERP is a subscriber for events involving material. The message is passed to the inventory module for handling. Inventory responds through the ERP portal to the bus where facility management is a subscriber to that information. Because the CIS is not interested in materials, it does not subscribe to inventory availability messages.

All rules for publishing and subscription are handled in the message broker, the common repository for all rules of connectivity, permissions, and relationships. This example includes a portal for Web service. This extends the scope of included applications to include distant suppliers or perhaps other separate agencies with data of interest (highway departments, assessors' offices, etc.). This concept remains unused or rare among water and wastewater utility organizations. However, it is being used within larger organizations in the energy and communication industries. Figure 8.7 shows an example of an integration based on publish and subscribe.

- Advantages. Adaptable to changing organizations. Can manage multiple suites.
- Disadvantages. Beyond needs of many utilities.

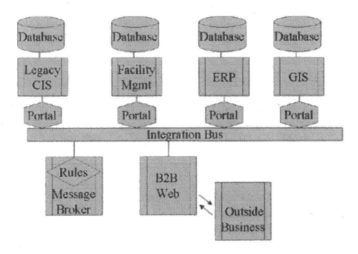

Figure 8.7   Example of an integration based on publish and subscribe.

REFERENCES

Azteca Systems (2002) Homepage: http://www.Azteca.com (accessed March 12, 2002).

Black, J. D. (2002) The Enterprise Melting Pot. GEOWorld http://geoplace.com/ge/1999/0499/499fea.asp (accessed February 18, 2002).

Booth, R.; Rogers, J. (2001) Using GIS Technology to Manage Infrastructure Capital Assets. *J. Am. Water Works Assoc.*, **November.**

City of Houston Public Works and Engineering. Resource Management/ Utility Customer Service/Automatic Meter Reading Program. http://www.ci. houston.tx.us/pwe/resource/usc/meterread.htm.

Doyle, M. J.; Rose, D. (2001) Protecting Your Assets. *Water Environ. Technol.*, **13** (7).

Ennis, J.; Boulos, P.; Heath, J.; Hauffen, P. Improved Water Distribution System Modeling and Management Using Map Objects. ESRI 2001 User Conference. http://www.esri.com/library/userconf/proc01/professional/papers/ pap169/pl169.htm (accessed November 20, 2001).

ESRI (2000) ArcFM Water: AM/FM/GIS for Water Utility Systems, an ESRI white paper; April.

ESRI (2000) ESRI and Oracle—Solutions for GIS and Spatial Data Management, an ESRI White Paper; August.

GEOWorld (2001) Interview with Peter M. Batty, GE Smallworld. http://www. geoplace.com/gw/2001/0112/interviewbatty.asp (accessed January 21, 2002).

GEOWorld (2001) Interview with Stewart Asbury, Byers Engineering Co. http://www.geoplace.com/gw/2001/0112/interviewasbury.asp (accessed February 18, 2002).

Government Accounting Standards Board (2004) http://www.gasb.org (accessed January 18, 2004).

Intergraph Corporation (2002) Utilities Solutions. http://www.Intergraph.com/Utilities (accessed March 12, 2002).

Joffe, B. (2002)How Good Are Your Maps? GEOWorld, http://www.geoplace.com/gw/2002/0202/0202dac.asp (accessed February 19, 2002).

MRO Software, Inc. (2002) http://www.MRO.com (accessed June 7, 2002).

Tibco Software, Inc. (2002) Homepage: http://www.tibco.com (accessed March 1, 2002).

Vitria Technology Incorporated (2002) Homepage: http://www.vitria.com (accessed March 1, 2002).

Water Environment Federation Collection Systems Committee (2004) http://www.cmom.net.

Wilson, J. D. (2002) GIS/ERP Integration Opens New Markets. GEOWorld. http://www.geoplace.com/gw/19980698busw.asp (accessed February 18, 2002).

Wilson, J. D. (1999) System Integrators Influence Geotech Projects. GEOWorld. http://www.geoplace.com/gw/1999/0299/299bwtch.asp (accessed February 18, 2002).

Chapter 9
Technology and Obsolescence Management

THE TECHNOLOGY LIFE CYCLE

Determining which geographic information systems (GIS) technologies to implement can be a daunting task for an organization. It is important to be on the leading edge of technology, but not the bleeding edge. The leading edge is dangerous, as all technology investments are long-term, and investing in unproven technologies can expose an organization to unnecessary risk. An example is the chip standards for modems. In the late 1990s, both Rockwell and U.S. Robotics were competing to create the next standard for modem chips. Those who invested in the Rockwell product lost heavily because U.S. Robotics won out.

Before decisions are made about selecting a new technology, difficult questions must be answered, including the following:

- What are the business processes that this technology will support?
- Will the technology meet the needs of the organization?
- Is the technology affordable?
- What are the software or hardware requirements?
- What amount of training will staff require, and will they accept the product?
- Will the technology be around for the foreseeable future?
- Who are the industry leaders providing the technology?
- What are the proprietary data, development, and platform issues?
- Will the technology interact smoothly with the organization's existing systems and data?

Answers to these questions may be found by gaining an understanding of the technology life cycle. This life cycle may be segmented into the stages of planning, budgeting, implementing, training, supporting, and evaluating as shown on Figure 9.1.

PLANNING. Technology planning may provide an organization with the tools it needs to break the cycle of constantly reacting to technology. Planning enables an organization to take proactive control of its technological tools and make them work for the organization. Building on open technology from strong vendors may reduce the risk of being stuck with obsolete or proprietary systems. Furthermore, a technology plan ensures that an organization uses a technology in ways that are sustainable, in addition to providing a clear understanding of the costs and requirements associated with technology implementation and management. Geographic information systems software vendors regularly provide new tools and capabilities so frequently that it can be disruptive and challenging to integrate each new product into an organization. Understanding how and determining when to introduce new technology will help an organization gain the most from any new technology.

Figure 9.1 Example of the technology life cycle.

BUDGETING. The costs of technology budgeting go far beyond the purchase price. An organization must be aware of the training, yearly maintenance, and upgrade costs of a technology. The general rule of thumb is that for every 30 cents spent on hardware and software, another 70 cents will likely be spent in training, support, and upgrades. That is, for every dollar spent on hardware and software, 2.3 times that amount should be budgeted for the ongoing cost of the technology.

IMPLEMENTATION. Technology implementation may range from an upgrade or revision of an existing system to full implementation of a new technology. The new technology solution could include a new software, hardware, application, database, or network system configuration. Once the new technology is in place, a series of tests must occur, with potential reconfiguration where necessary.

TRAINING. Training for staff who will use the new technology should be viewed as a critical success factor in the technology life cycle. Training allows the staff to get comfortable and capable with the new technology. Training may range from simple Web-based education or a CD-ROM tutorial, to a training seminar or on-site training by a qualified educator.

SUPPORT. To minimize problems, GIS staff should spend their time supporting the new technology by tracking upgrades and patches, providing backup of all software and data, and maintaining documentation.

EVALUATION. An evaluation of a new technology, once implemented, should be undertaken to measure its success. There are two criteria to consider when determining if the technology meets your needs. The first is to determine if the technology meets the expectations of its users as measured by reviewing the initial design documents and noting if users are using it. The second is to determine if the technology helps improve an organization's ability to achieve its mission (NPower, 2002).

TECHNOLOGY OBSOLESCENCE. Information technology changes very quickly; therefore, one of the more critical decisions an organization must make is to select technology that will not be obsolete within a short period. Once the investment is made, additional management considerations must be taken into account to ensure that the investment is maximized. One of the first considerations should be that all information technology should be standardized within an organization.

Standardization should be considered for hardware, including computers, backup devices, and peripherals, because once an investment is made, it will need to be supported. Components produced by a single manufacturer typically will perform together better than a hybrid system in the long run as technologies evolve and new functions and components are added. With a standardized system, it is generally easier to find and incorporate compatible devices to achieve a system upgrade. Thus, building a system around standardized hardware should help to forestall obsolescence. An organization should select the minimum array of standardized hardware that meets the organization's information technology requirements within the available budget. Hardware standardization ensures compatibility among all hardware devices, reduces the number of devices to support and the resources required to support it, and greatly simplifies the support requirements and associated level of knowledge and expertise necessary to provide effective support. It also allows information to be shared more dependably and easily between departments or divisions, reducing the likelihood of any one group being isolated. Interoperability between departments in an organization will provide better interdepartmental integration and efficiency.

A reason to consider upgrading might include an exponential leap in technology capability or functionality, as in the case of CD-ROM versus DVD technology. A CD-ROM used as a storage device can typically hold 650 megabytes of data. A DVD can hold as much as 5.8 gigabytes—an order of magnitude more than CD-ROM. While the CD-ROM is standardized in the industry, DVD has competing standards, so the risk exists of investing in a DVD format that may ultimately not become an industry standard. Therefore, in deciding whether to embrace a new technology, such as DVD, an organization should consider how information will be shared among internal and external users and whether users will be able to readily use it.

Another example of a critical decision process involving technology upgrades is when to upgrade software to new versions. Typically, new versions will offer some additional functionality that may be appealing but may require additional staff training and support. New software versions may also be less stable initially, resulting in lost data or time. Management consideration should be given to adopting new versions once the software demonstrates stability.

THE VALUE OF DATA

Data is one of the single greatest investments an organization will make. It is estimated that as much as 80% of an organization's budget will be spent on the initial development and maintenance of data. Applications and technology that use data in different ways may change; however, the needs to manage data and ensure the quality of data will remain.

Managing data cannot be wholly separated from the need to manage technology to ensure long-term viability and prevent or minimize obsolescence. Information that is supported by obsolete technologies can become inaccessible after a technology upgrade. Digital storage media and the data in them often are needed long after the technology that supports it is replaced. Thus, data management and maintenance can involve translating or migrating data to new formats as technologies evolve. Data standards and metadata, discussed below, can greatly facilitate effective and efficient data migration.

Maintenance of the data is an ongoing task throughout the life of the GIS and must be recognized as an institutional investment and asset. To ensure this, communicating the value of data to an organization's leadership is critical and should be managed as an organizational asset.

WHY DATA IS IMPORTANT. Good data is required for analyses and to create maps using GIS. For example, to reveal relationships in water consumption information, one would use billing information and the customer addresses database. One would not only need to ensure that those addresses are correct for the map to be useful, but that the consumption data are current.

DATA TYPES AND MODELS. Data used in GIS is typically one of the following three basic forms:

- Spatial data. Geographic features referenced to real-world coordinates and made up of points, lines, and polygons. Spatial data forms the locations and shapes of map features, such as parcels, streets, or utility systems.
- Tabular data. The nongraphic attributes stored in a relational database that describe each geographic feature.
- Image data. Satellite images, aerial photographs, and scanned data.

In addition, these data can be further classified into two types of data models, raster and vector data, which are described below.

Spatial Data. Spatial data include points, lines, and polygons and are the typical components of a map that most people think of. They are defined as follows:

- Points represent anything that can be described in terms of an *x, y* coordinate location on the face of the earth, such as wells, manholes, poles, and street signs.
- Lines represent anything having a length, such as roads, streams, pipelines, and water mains.
- Polygons or areas describe anything having boundaries, whether natural, political, or administrative, such as the boundaries of countries, states, political entities, buildings, parcels, and lakes.

The spatial data described above represents one of two models used to represent data in a GIS as illustrated on Figure 9.2. It is called a vector model. In a vector model, each feature is defined by *x, y* location coordinates in space (the GIS connects the dots to draw lines and outlines, creating lines and areas). Another model is the raster model in which features are represented as a matrix of cells in continuous space. A point is one cell, a line is a continuous row of cells, and an area is represented as continuous touching cells (Dictionary of GIS Terminology, 2001).

VECTOR FORMATS. Much of GIS data is based on vector technology, so vector formats are the most common. There are many ways to store coordinates, attributes, attribute linkages, database structures, and information display, resulting in a much more complex format. Some of the more common formats are briefly described in Table 9.1.

RASTER FORMATS. Raster files generally are used to store image information, such as scanned paper maps or aerial photographs. They are also used for data captured by satellite and other airborne imaging systems. Images

- **Raster or grid GIS**

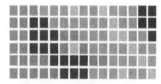

- **Vector GIS (points, lines, and polygons)**

Figure 9.2   Raster versus vector data.

Table 9.1 Common vector file formats.

Format name	Software platform	Internal or transfer	Developer	Comments
Arc Export	ARC/INFO[a]	Transfer	ESRI	Transfers data across ARC/INFO[a] platforms.
ARC/INFO[a] Coverages	ARC/INFO[a]	Internal	ESRI	
AutoCAD Drawing Files (DWG)	AutoCAD[a]	Internal	Autodesk	
Autodesk Data Interchange File (DXF™)	Many	Transfer	Autodesk	Widely used graphics transfer standard.
Digital Line Graphs (DLG)	Many	Transfer	U.S. Geological Survey (USGS)	Used to publish USGS digital maps.
Hewlett-Packard Graphic Language (HPGL)	Many	Internal	Hewlett-Packard	Used to control HP plotters.
MapInfo Data Transfer Files (MIF/MID)	MapInfo[a]	Transfer	MapInfo Corp.	
MapInfo Map Files	MapInfo[a]	Internal	MapInfo Corp.	
MicroStation Design Files (DGN)	MicroStation[a]	Internal	Bentley Systems, Inc.	
Spatial Data Transfer System (SDTS)	Many (in the future)	Transfer	U.S. Government	New U.S. standard for vector and raster geographic data.
Topologically Integrated Geographic Encoding and Referencing (TIGER)	Many	Transfer	U.S. Census Bureau	Used to publish U.S. Census Bureau maps.
Vector Product Format (VPF)	Military mapping systems	Both	U.S. Defense Mapping Agency	Used to publish Digital Chart of the World.

[a]Source: http://www.gisdevelopment.net/tutorials/tuman003.htm.

from these systems are often referred to as remote-sensing data. Unlike other raster files, which express resolution in terms of cell size and dots per inch (dpi), resolution in remotely sensed images is expressed in meters, which indicates the size of the ground area covered by each cell. Some of the more common raster formats are briefly described in Table 9.2.

Tabular Data. Tabular data consist of the attribute information that describes spatial features. Tabular data can also include lists, spreadsheets, or databases containing information such as customer lists, restaurant locations, or inspection data. Tabular data, as shown in Figure 9.3, can be linked with spatial data in the GIS to visualize more information than location.

Image Data. Images can be used as base maps and can be displayed along with other spatial data containing map features. Examples are aerial photography, satellite imagery, or digital photos. Using image data can often be more cost- and time-effective than trying to collect the corresponding data layers individually, especially for large areas. However, one drawback is that image data is one file, or layer, and cannot be broken into the different components readily. Image data is the best choice if one needs to add a point of reference to vector data without attaching additional information. Images can also be attributes of map features. In other words, images can be added to other map features so that clicking on the feature would display the image. Almost any document or photograph can be scanned and stored as an attribute in a GIS database that can be displayed within the GIS (Dictionary of GIS Terminology, 2001).

DATA MODELS. A geographic data model is basically an abstraction of the real world, defined as spatially located features with attributes. A data model is designed to best manage entities and the relationships between those entities. The physical description of the world is diverse, but can be divided into five broad categories for GIS applications.

- Features. Discrete objects on a map represented by points, lines, and polygons.
- Surfaces. Two-dimensional triangulated irregular network (TIN) or grid representations of surfaces, such as the earth's surface or geophysical phenomena.
- Network. A set of features in a linear system, such as a utility network or transportation network.
- Image. Raster-based data representations of captured imagery of the planetary surface and its environment.
- Location. Location-based information related to addressing, postal zip code areas, or survey control locations.

Computer-Aided Design Data Model. The first computerized mapping systems used vector-based map technology of computer-aided design (CAD) and the CAD data model to describe features of the world as points, lines, and

---

Table 9.2. Common raster file formats. [a]

Format name	Software platform	Internal or transfer	Developer	Comments
Arc Digitized Raster Graphics (ADRG)	Military mapping systems	Both	U.S. Defense Mapping Agency	
Band Interleaved by Line (BIL)	Many	Both	Common remote-sensing standard	
Band Interleaved by Pixel (BIP)	Many	Both	Common remote-sensing standard	
Band Sequential (BSQ)	Many	Both	Common remote-sensing standard	
Digital Elevation Model for (DEM)	Many	Transfer	U.S. Geological Survey (USGS)	USGS standard format digital terrain models.
PC Paintbrush Exchange (PCX)	PC Paintbrush	Both	Zsoft	Widely used raster format.
Mr. SID	Many	Internal	LizardTech	Widely used compression format that is popular for its data storage benefits.
Spatial Data Transfer Standard (SDTS)	Many (in the future)	Transfer	U.S. Federal Government	New U.S. standard for both raster and vector geographic data; raster version still under development.
Tagged Image File Format (TIFF)	PageMaker	Both	Aldus	Widely used raster format.

[a]Source: http://www.gisdevelopment.net/tutorials/tuman003.htm

Shape	Graphic_id	Starting_at	Length	Adj_up_km	Adj_dn_km	Graph_km	Flowtype	Pipe_cost	Ring_dia	Mouth	Upstream	Downstream	ds_full	Primary_id
PolyLine	1091	1091	297.63621	317.380	311.730	297.05296	GR			8.000	1007	402160	402196		6513
PolyLine	1089	1089	158.50208	311.730	308.730	159.51772	GR			8.000	1007	402156	402154		6513
PolyLine	1088	1088	302.01127	308.730	286.690	303.14697	GR			8.000	1007	402154	402148		6513
PolyLine	1085	1085	394.68277	297.610	264.090	394.86343	GR			10.000	1007	402143	402133		6513
PolyLine	1086	1086	314.75414	286.690	253.380	316.37016	GR			8.000	1007	402148	402153		6513
PolyLine	1084	1084	162.41529	284.090	275.930	162.45909	GR			10.000	1007	402133	402115		6513
PolyLine	1083	1083	184.08224	275.930	273.230	184.09847	GR			10.000	1007	402115	402112		6513
PolyLine	1082	1082	464.36982	273.230	267.250	464.40232	GR			10.000	1007	402112	402090		6513
PolyLine	1081	1081	257.61986	267.250	247.230	257.97574	GR			10.000	1007	402090	402083		6513
PolyLine	1087	1087	213.31218	253.390	213.800	215.93430	GR			8.000	1007	402153	402159		6513
PolyLine	2534	2534	273.88893	249.580	246.530	375.30464	GR			12.000	1007	602181	602180	1031	6513
PolyLine	1080	1080	234.54026	247.230	222.590	235.49236	GR			10.000	1007	402083	402071		6513
PolyLine	2535	2535	269.88820	246.530	243.300	269.88820	GR	VCP		12.000	1007	602180	601898	1031	6513
PolyLine	2536	2536	364.16745	243.230	240.090	364.16745	GR	VCP		12.000	1007	601898	601507	1031	6513
PolyLine	2249	2249	398.03596	243.200	240.860	398.03860	GR			8.000	1007	401272	401273		6513
PolyLine	1039	1039	368.83673	240.860	238.590	368.85016	GR			8.000	1007	401272	401293		6513
PolyLine	2537	2537	213.94196	239.930	234.860	213.94196	GR	VCP		12.000	1007	601507	601506	1031	6513
PolyLine	1040	1040	266.64441	238.590	231.830	266.85067	GR			8.000	1007	401293	401345		6513
PolyLine	2538	2538	457.15612	234.760	214.450	457.15612	GR	VCP		12.000	1007	601506	601441	1031	6513

Figure 9.3   GIS tabular data.

polygons and store them in separate files on a computer. Attribute information was scanty, at best, with annotation and layers as the primary modes of representation.

Geographic Information Systems File-Based Model. Beginning as late as the 1970s, vendor and organization-specific data formats (such as the Environmental Systems Research Institute's [ESRI] coverage data model) addressed the need to support the display, querying, editing, and analyses of spatially related information and its attributes in some graphical manner. These new data models ensured that attribute information was combined with spatial information, and that the topological relationships between spatial features were stored. For example, a utility pipeline's diameter and material attribute data could be displayed and the upstream and downstream pipe segment IDs. What was lacking was the ability to associate behavior with a particular map feature.

Spatial Database Model. There are now many software products that support the storing of spatial data features. These data can be accessed by a further list of GIS software products. An example is the ESRI geodatabase. The geodatabase concept allows for storage and transfer of spatial and attribute data on a personal geodatabase (i.e., Microsoft Access) to client-server versions (i.e., Oracle, SQL Server, Informix). Access is through any of ESRI's GIS software products and CAD-client capability.

DATA STANDARDS. Through the efforts of the Federal Geographic Data Committee (FGDC) and the Computer-Aided Design and Drafting (CADD)/ GIS Technology Center, there is now an effective and increasingly widely used set of standards for describing data. These standards, among others, now allow users to search through terabytes of data now offered by public and private organizations (Goodchild, 2001).

Data standards offer a variety of specifications for the many ways data is described. Standards encompass the specification of data file and field names, how the data are organized, and how and in what structure data is described (metadata). Data standards facilitate data sharing, enforce and track data quality, and have a long-term effect on shaping the way geospatial technology and services are delivered. There are several sources of geospatial data and technology standards.

- De facto standards. While not a standard in terms of being developed by a recognized standards organization, these become standard through use by the vendor and user communities. The ESRI shapefile is a good example.
- Government standards. Typically developed by a federal government agency. Examples include the Spatial Data Standard for Facilities, Infrastructure, and the Environment (SDSFIE) data standard and the FGDC metadata standard. These government standards support the many needs of data providers and users of published government data.
- Formal standards. Standards and specifications developed through a formal process managed by a recognized standards organization.

The ways in which all of these formats are used within each organization need to be reviewed periodically to prevent obsolescence. Some data formats might become obsolete as the technology evolves and will be lost if it is not translated to a more updated format or standard.

Leaders in developing government standards are the FGDC, an arm of the U.S. Geological Survey at www.fgdc.gov, and the CADD/GIS Technology Center's SDSFIE at www.tsc.wes.army.mil. These agencies have made great progress in developing data standards that have been adopted by the geospatial community.

The FGDC standards are intended to provide guidance and direction to FGDC standards developers and users. FGDC standards outline the responsibilities of federal agencies with respect to coordinating federal surveying, mapping, and related spatial data activities. The purposes are to develop a national spatial data information resource, reduce duplication, reduce the expense of data collection, and facilitate the sharing of available data.

Executive Order Number 12906, April 1994, designates the FGDC as the lead entity to coordinate the National Spatial Data Infrastructure (NSDI), which is defined as the technology, policies, standards, and human resources necessary to acquire, process, store, distribute, and improve utilization of geospatial data. There are many standards that may support the NSDI, some of which are listed in Table 9.3 (Federal Geographic Data Committee, 1996).

The CADD/GIS Technology Center was chartered to coordinate the development of SDSFIE, which provides capabilities and satisfies needs for CADD, GIS, and computer-aided facility management (CAFM) technology applications throughout the U.S. Department of Defense (DoD); federal, state, and local governments; and the private sector. This includes setting standards, promoting system integration, supporting centralized acquisition, and providing assistance for the installation, training, operation, and maintenance of CADD/GIS and facilities management (FM) systems. This also includes directing specific application developments, promoting communications, developing and promoting standards, furnishing technical advice, coordinating with professional

Table 9.3   Various standards supporting the national spatial data infrastructure (FGDC, 1996).

Standard type	Description
Agency standards	Agency standards may be developed to support specific applications or mandates within one agency. Any agency or organization may support or recognize an agency standard. Typically, the sphere of influence in the development, maintenance, and use of an agency standard is contained within a single agency. Cooperative agreements between or among agencies to develop specific standards are included in agency standards.
Federal Information Processing System standards	Federal Information Processing System standards are developed to standardize data and processes among federal agencies. Their goal is to gain efficiency and economy through widespread use. These standards are generally mandated for use by federal agencies.
FGDC Standards	FGDC Standards are developed in response to OMB Circular A-16 and EO 12906, which mandate data sharing and adherence to common standards for federal agencies. They are intended to be national in scope and to go beyond individual agencies and the federal government enterprise. They support national and collective decision making and applications and are developed jointly by federal, state, and local governments and other interested participants. They are only mandatory for federal agencies.
Industry standards	Industry standards are developed in the private sector by cooperating firms. Their production may be coordinated by a single firm, a group of firms, a not-for-profit organization, or a standards organization. These standards are voluntary unless conformance is mandated through contract or agreement.
American National Standards	American National Standards (ANS) are endorsed by the American National Standards Institute (ANSI) and are national in scope. These are voluntary standards developed and supported by commercial industries that implement technology, but any individual or organization, including governments, can participate in the development of an ANS.
International standards	The International Organization for Standardization (ISO) is the primary international standards organization for information technology. Organizations gain access to ISO through their national standards body. In the U.S., this is ANSI. Federal Geographic Data Committee standards may be affected in their development or adoption by other standards or may affect or contribute to other standards.

organizations and industry, evaluating technological developments, and recommending necessary CADD/GIS and FM policies to ensure that the maximum benefits are realized from these technologies.

The SDSFIE is the only nonproprietary GIS standard designed for use with the predominant commercially available off-the-shelf GIS and CADD applications (e.g., ESRI ArcInfo and ArcView; Intergraph MGE and GeoMedia; AutoDesk AutoCAD, Map and World; and Bentley MicroStation and Geo Graphics), and relational database software (e.g., Oracle and Microsoft Access). This nonproprietary design, in conjunction with its universal coverage, has propelled the spatial data standards (SDS) into becoming the standard for GIS implementations throughout the DoD, as well as the de facto standard for GIS implementations in other federal, state, and local government organizations; public utilities; and private industry throughout the United States and the world (CADD/GIS Technology Center, 2002).

USING GEOGRAPHIC INFORMATION SYSTEMS DATA. It is estimated that approximately 80% of all data has a spatial component; data from most sciences can be analyzed spatially. Information that is needed by any organization is stored in a GIS as a collection of data layers containing similar data that can be used together. A layer can be anything that contains similar features, such as parcels, streets or roads, buildings, streams and lakes, or utility systems. Each layer requires some geographic reference, such as a coordinate value (i.e., latitude and longitude), or some other location reference, such as a postal code, census tract, or address, to be used in GIS.

Many different types of data can be used within a GIS, including a wide range of proprietary and standard map and graphics file formats, images, CAD files, spreadsheets, relational databases, and data from many other sources. Data are free or fee-based and come from commercial, nonprofit, educational, and governmental sources; other GIS software users; and one's own organizations (Dictionary of GIS Terminology, 2001).

Selecting the Right Data. Determining the types and formats of data that are right for an organization is typically driven by the business processes one is trying to address and the organization's priorities. Following are some of the more important questions that need to be considered when determining which data is needed.

(1) What does one want to do with the data?
(2) What are the specific geographic features needed?
(3) What attributes of those features are needed?
(4) What is the geographic extent of the area of interest?
(5) What is the level of geography one wants to examine within the area of interest?
(6) How current must the data be?
(7) What type of computing environment will be used?
(8) What GIS software will be used?

(9) How many concurrent users will be accessing the data, at how many locations?

(10) When is the data needed?

(11) Will periodic data updates be needed and, if so, how frequently?

(12) Which of the data sets identified may be licensed from the same data publisher?

(13) Does one plan to start small, then expand?

(14) Does one want to publish derivative products with the data?

Sources. Organizations that use GIS often integrate data from numerous sources, such as aerial photographs and as-built drawings, maintenance and document management system databases, and land base and utility maps. Data used in GIS applications can come from a variety of sources, such as

- Digitized hard copy maps;
- Scanned documents;
- Databases, spreadsheets, and American Standard Code for Information Interchange (ASCII) data files;
- Field data collection such as global positioning system, personal digital assistant (PDA), mobile, and inspection data;
- Remote sensing;
- Aerial photography;
- Internet data sources; and
- Other organizations and agencies.

In addition, some significant questions about the data need to be asked (i.e., what is available, where is it located, how current is it, and how accurate is it?). Whatever the source or quality, success of GIS depends on using as much relevant data as possible. These data integration issues can be addressed in a number of ways, each having associated costs and effects to the enterprise. Some of the solutions are

- Translation programs, which regularly replicate data into new formats;
- Enterprise application integration initiatives, which reengineer processes; and
- One-time conversion projects, which enable new data maintenance environments (Dictionary of GIS Terminology, 2001).

Data Ownership. As mentioned at the beginning of the chapter, data is one of the single greatest investments an organization will make. As such, it must be protected and its use restricted to maintain its value to the organization and protect privacy.

Data Piracy. The data that an organization "owns" or pays to create can easily be pirated for use by any individual with a CD-RW drive or access to your data via the Internet. Recent law concerning data piracy focuses on the 1991

U.S. Supreme Court case, *Feist v. Rural Telephone Service Co., Inc*. This case determined that "If you own a geospatial product . . . people can extract and copy it to their heart's content. They can even use it to compete with you." (Tech Law Journal, 2002). Data protection laws are complex, but there are three important exceptions to this ruling.

- Arrangements of data. Owners can still protect the way they arrange their data if they use minimal creativity. This is a strategy for many geospatial products.
- Software. Most geospatial products are useless without advanced search tools. Unlike data, software is copyrightable.
- Contracts. *Feist* did not change state contract law. Vendors can still write contracts that restrict users' ability to copy and redistribute data.

It must be noted that the data protection industry is still in its infancy. Many of the rules that govern market actions have yet to be written. New congressional legislation may create "collections of fact" that are hidden from public view, or allow monopolies to flourish, creating noncompetitive, fix-priced markets for their product. Such tinkering could have significant drawbacks for the geospatial community (Maurer, 2002).

Freedom of Information Act. The Freedom of Information Act (FOIA) of 1966 was enacted by Congress to give public access to information held by the federal government. It gives any person the right to request and receive any document, file, or record in the possession of any agency of the federal government. There are exceptions to the FOIA, such as data and classified materials related to national security, trade secrets, personnel, medical files, and even geological and geophysical data and maps concerning wells.

Because of the FOIA, the Electronic Freedom of Information Act Amendments, and many states' open record policies, government-held data are owned by the public. Such data include a water utility's customer address water usage, and phone log databases to sensitive property or voting records. Geographic information systems data being put on the Internet has caused citizens to be increasingly concerned about privacy. For an organization that has generated or maintains data susceptible to such regulations, a balance must be made between responsible data publishing and protection of a citizen's privacy. The GIS staff of an organization must make themselves aware of public opinion concerning published information and be governed by ethical and professional standards. Use of software solutions might provide some valuable tracking and security tools for managing FOIA-susceptible data (Sadagopan et al., 2000).

Data Maintenance. Data must be maintained to ensure its quality and usefulness in supporting the mission of the organization. Maintaining data to ensure its accuracy, currency, and that it is in a format that is not faced with technological obsolescence is, therefore, crucial to an organization's mission. First, make note of the initial data quality by generating a set of metadata (see next section) for each data set. All subsequent data revisions must be noted, such as changes to the projection of a data set or realignment of a utility line based on more accurate survey information. Simple updates to specific attribute data records need not be tracked in the metadata; rather it is the addition of a field or links to other external information solutions resources that need be noted.

Next, ensure that the organization's data are backed up regularly using tape, CD-ROM, CD-RW, or DVD media technologies. The life spans of these media range from 5 years for 8mm tape to 10 years for CD or DVD technologies. To plan for a long-term archive or backup of data, make sure that the hardware and software interfaces selected have longevity.

METADATA. *Metadata* can be defined as the data about the data. Metadata can provide a description of a data set, its quality, its depth, the date it was generated, and its aerial coverage. A map legend is a perfect example of metadata because it contains information on the originator of the data, the date it was published, a description, the map scale and data accuracy, the referenced coordinate system and datum, and other valuable information. The end user will have a certain level of confidence in the map data based on this information (Federal Geographic Data Committee, 1994). Metadata assists an organization in managing its data by capturing information that would otherwise have been lost because of employee turnover, or that would be lost or forgotten by staff over time. For end-users, metadata provides a tool enabling them to find the best data for their needs. Metadata also makes data-sharing activities between organizations possible. Hence, this information should be viewed as a critical part of any data set and a critical success factor for any GIS group.

Metadata is typically under-valued by an organization. Geographic information systems staff typically balk at the effort involved in metadata upkeep, but miss the value it brings. To motivate GIS staff, management must communicate the benefits that metadata entry brings, such as order, standardization, quality control and quality assurance, autonomy, identity, and a sense of achievement.

There are two general classes of metadata standards. Homegrown metadata standards are those developed by the data originator or by the managing organization. These standards, though meeting the needs of that particular set of data users, do not enable data sharing nor account for the needs of other potential data users. Government metadata standards are those developed by a federal government agency. The clear leaders in this effort are agencies such as the FGDC at www.fgdc.gov and the CADD/GIS Technology Center's SDSFIE at tsc.wes.army.mil. Other valuable information can be found at www.tenlinks.com/MapGIS/reference/metadata.htm.

To ensure that metadata entry occurs, an organization should implement long-term metadata entry strategies, such as

- Link metadata entry directly to GIS staff work duties. Accountability for metadata population and management, with a reward system for success, should create the proper culture needed for metadata entry.
- Organization and GIS management should be proponents of metadata standards and ensure that, as staff turnover occurs, new staff are fully trained in metadata culture and techniques.

- Provide tools and techniques that assist in metadata capture and management. Also find more creative uses for this valuable set of data.
- Provide peer pressure to those organizations that do not have their own metadata documentation, or who create their own documentation methods that do not conform with already existing standards, such as the FGDC data standards (Mathys, 1999).

There are now a variety of software tools available to assist with metadata collection. Some of these tools are stand-alone products, while others come bundled with a GIS software product. Also, many GIS data sets now come with associated metadata, typically in the FGDC format (ESRI data sets, for example). For a more complete listing, see www.badger.state.wi.us/agencies/wlib/sco/metatool/mtools.htm.

In the future, metadata will also play a more important role in initiatives being implemented by the Open GIS Consortium. Metadata will provide the ability to do real-time projections and file conversion through standards being implemented. This will allow the representation of any data format in any coordinate system. Metadata will be a key to successfully implementing standards.

*M*ANAGING VENDORS

It is important to recognize that the vendor's motives are not the same as the client's. The client is ultimately responsible for purchasing a product that will meet the needs of the organization for years to come. The vendor wants to sell a product. When the client interacts with a vendor, recognize that it is a negotiation. The client must thoroughly understand the business processes of his or her organization and approach the software vendor with knowledge of these needs. Basic questions like "Can the product do X while interfacing with Y data?" should be thought out in advance. The vendor then is put in the position of meeting the client's needs one-by-one. This process could take several meetings.

Long-term vendor relationships are critical. When a vendor begins to decline in the market, the client is left unsupported and needs to identify an alternative product. An example is the MicroStation GIS Environment (MGE) product line, which has limited research, development, and enhancements. Institutions that invested in MGE must now realize the effect of their investment and budget for a shift in technologies or products.

Partnerships between an organization and a software vendor have their own risks. They typically do not work if the vendor has the upper hand in the relationship or if a personal relationship is not built between the organization and the vendor. Such situations leave the organization at the mercy of the vendor. Yet, vendors typically appreciate and need market feedback to enhance their products. This is the only way organizations will continue to use their products and recommend them to others.

INDUSTRY AND USER GROUPS

Using and participating in software user and industry groups can provide users with several benefits. First, formal interaction with others who are interested in similar issues and similar data can help one understand how others are using information that one might be interested in, sources of data, and opportunities for data collaboration. Second, industry and user groups are typically formed to allow GIS users a chance to share ideas, resources, and data, and to act as a vehicle for gaining a better understanding of GIS and related technologies. Finally, industry and user groups may have a better chance of implementing change in GIS software from vendors. By having multiple organizations interested in having additional functionality added to the GIS, the enhancements will have a better chance of being implemented.

THE IMPORTANCE OF INDUSTRY AND USER GROUPS. Local or regional industry and user groups are extremely useful for providing a forum for persons interested in GIS and related technologies to share ideas, resources, and data. They also create an atmosphere of collaboration and cooperation. This environment creates a virtual support organization of software and data development experts who are spread among multiple organizations. Any organization can now leverage these other organizations for information regarding such tasks as data collection and training.

Industry groups facilitate two-way communication between software vendors and an organization. Through the industry group, an organization can influence the creation of certain standards in data and software development. For example, an organization could influence how hydraulic modeling software is created, water flow is modeled, and billing for the product occurs. The industry group, in turn, should be able to influence the client by communicating what is becoming de facto technology, where the industry is going, and which other agencies are using a particular software solution or data standard.

WHERE TO LOOK FOR INDUSTRY AND USER GROUPS. There are many GIS industry and user groups formed in most of the metropolitan areas of the United States. These groups are formed by industry and organizations (i.e., urban data experts, natural resources information repositories, utilities, and universities), by geographic location (country, regional, statewide, or local), and by software product (ESRI, MGE, and MapInfo user groups). Any search on the World Wide Web will produce a near-complete list of such groups. A listing of some of the major industry and user groups is provided below.

Resources.

- GIS Café—www.giscafe.com/USERSGROUPS/, a Web site acting as a portal for GIS news and resources.

- Geo Community—http://careers.geocomm.com/usergroups/, a Web site for GIS, CAD, mapping, and location-based industry professionals, enthusiasts, and students to interact.

Vendors.

- Autodesk User Group International—http://www.augi.com, Autodesk's officially recognized global organization, which creates and maintains networking opportunities, educational programs, global communication channels, and opportunities for user involvement.
- ESRI—http://www.gis.esri.com/usersupport/usergroups/usergroups.cfm. The Environmental Systems Research Institute has several user groups, and specialty industry groups exist independently of ESRI and meet throughout the year to share information, data, tips on software usage, and project news.
- Intergraph—www.intergraph.com/gis/community. The Intergraph GeoSpatial Users Community (IGUC) is dedicated to serving the needs of users in the mapping and GIS industry.
- MapInfo—www.mapinfo.com/support/user_groups. MapInfo presently has many user groups in the United States and around the world. These groups provide a perfect environment for MapInfo users to help each other with technical tips; share knowledge about applications, data, and services; and socialize with professionals in widely varying fields.

Associations and National Groups.

- Association of American Geographers (AAG)—http://www.aag.org. The AAG is a scientific and educational society founded in 1904. Its 6500 members share interests in the theory, methods, and practice of geography, which they cultivate through the AAG's Annual Meeting, two scholarly journals (the *Annals of the Association of American Geographers* and *The Professional Geographer*), the monthly *AAG Newsletter*, and the activities of its two affinity groups, nine regional divisions, and 53 specialty groups.
- Canadian Association of Geographers (CAG)—www.zeus.uwindsor.ca/cag/cagindex.html. The CAG is the only national organization in Canada representing practicing geographers. The CAG organizes the principal annual conference on the discipline of geography, welcoming papers, panel discussions, workshops, and field trips presented by members.
- FGDC—www.fgdc.gov. The Federal Geographic Data Committee coordinates the development of the National Spatial Data Infrastructure (NSDI). The NSDI encompasses policies, standards, and procedures for organizations to cooperatively produce and share geographic data. The FGDC develops geospatial data standards for implementing the NSDI in consultation and cooperation with state; local, and tribal governments; the private sector and academic community; and, to the extent feasible, the international community.
- GITA—www.gita.org. In 1998, AM/FM International changed its name to the Geospatial Information and Technology Association (GITA) to

better reflect the Association's new focus. The GITA's mission is to provide excellence in education and information exchange on the use and benefits of geospatial information and technology in telecommunications, infrastructure, and utility applications worldwide.

- National States Geographic Information Council (NSGIC)—www.nsgic.org/indexframe.html. The NSGIC is an organization of states committed to efficient and effective government through the prudent adoption of information technology. Members of NSGIC include delegations of senior state geographic information system managers from across the United States.
- Open GIS Consortium (OGC)—www.opengis.org. The OGC is an international industry consortium of more than 220 companies, government agencies, and universities participating in a consensus process to develop publicly available geoprocessing specifications. The OGC provides many opportunities for participation. Visitors can read and comment on open GIS discussion papers and explore opportunities to submit proposals for open GIS specifications or interoperability initiatives.
- URISA—www.urisa.org. The Urban and Regional Information Systems Association (URISA) is a nonprofit association of professionals using GIS and other information technologies to address challenges in all state and local government agencies and departments.

HOW TO FORM AN INDUSTRY OR USER GROUP. If a person has determined that no industry or user groups are located near them, then forming such a group on one's own can be a rewarding experience. The steps to form such a group include the following:

- First, contact any existing groups;
- Talk to others in the industry who might be interested in forming a group;
- Identify a neutral location to hold meetings;
- Determine a reasonable meeting time; and
- Determine how the group should be facilitated.

*R*EFERENCES

CADD/GIS Technology Center (2002) *Spatial Data Standard for Facilities, Infrastructure, and the Environment.* http://www.tsc.wes.army.mil (accessed March 2002).

Dictionary of GIS Terminology (2001) The ESRI Press: http://www.gis.com; April.

Federal Geographic Data Committee (1994) *Metadata or "Data about Data".* U.S. Geological Survey: Reston, Virginia.

Federal Geographic Data Committee (1996) *FGDC Standards Reference Model.* U.S. Geological Survey: Reston, Virginia.

Goodchild, M. F. (2001) Finding GIServices. *Geospatial Solutions*, **January.**

Mathys, T. (1999) The Minnesota Metadata Missions. *Geo. Info. Systems*, **November.**

Maurer, S. M. (2002) Protecting What's Yours. *Geo. Info. Systems*, **February.**

NPower (2002) The Technology Product Lifecycle. http://www.npower.org. (accessed March 2002).

Sadagopan, G. D.; Richardson, J. J.; Singh, R. (2000) Complying with the Freedom of Information Act. *Geo. Info. Systems*, **January.**

Tech Law Journal (2002) http://www.techlawjournal.com/cong106/database/19910327feist.htm.

Index

A

Actors, 130–131
Advantages, 158
Advantages and disadvantages of
    geographic information system
    (GIS) applications, 114–117
  advantages, 114–116
  disadvantages, 116–117
American Standard Code for Information
    Interchange (ASCII) data files,
    224
Annals of the Association of American
    Geographers, 229
Application design, 121–133
  additional requirements, 131–133
    data concurrency, 132
    diversity of client platforms, 132
    network infrastructure, 133
    performance requirements, 132
  the application life-cycle, 121–128
    deployment, 127
    final development, 126–127
      business rules layer (analysis
        layer), 27
      data layer, 126–127
      presentation layer, 127
    final specification, 126
    maintenance, 127–128
    needs requirements, 122–123
    prototype development, 126
    prototype review, 126
    solution domain, 124–126
    specifications, 123–124
    testing, 127
  object-orientated design concepts,
    128–130
  unified modeling language, 130–131
Application development process, 118–121
Arc8, 106
ArcGIS, 121, 156
ArcInfo, 106, 119–121,223
Arc Macro Language (AML), 119
ArcObjects, 121
Arcs, 2, 32
ArcTools, 120–121
ArcView, 113, 119–120, 223
Attribute entry, 85
Audio Video Interleaved (AVI), 120
AutoCAD, 223
Autodesk's AutoLISP, 119, 223
Automated mapping (AM), 3–4
Automated mapping/facilities manage-
    ment (AM/FM), 4, 7–8, 19
Avenues, 119–120

B

Background, 150–151
Benefits derived, 32–40

H

Hardware maintenance, 159
Hydrologic and hydraulic (H&H)
models, 111, 119

I

Implement the plan, 64–74
  document outline, 72–74
  the implementation plan document, 72
  the planning process, 64–72
    related steps, 72
    step 1: perform needs analysis, 65–68
      compile needs, 68
      evaluate existing information
        systems, 67
      evaluating existing paper/com-
        puter-aided design (CAD)/geo-
        graphic information systems
        (GIS) data, 66–67
      initiate the process, 65
      interview participants, 65–66
      meet to review the needs analysis
        report, 68
      preparation, 65
      prepare draft needs analysis report,
        68
      update needs analysis report, 68
    step 2: define project vision, 68–71
      conducting vision workshop,
        69–71
      evaluate technology alternatives,
        69
      prepare GIS functional require-
        ments analysis, 68–69
    step 3: develop the implementation
      plan, 71–72
      define GIS data needs, 71
      develop implementation strategy,
        71
      develop system architecture, 71
      prepare final report, 71–72
      prepare preliminary report, 71
      present to executive staff, 72
      review the draft report, 71
Improvisational model of change
  management, 163–164
Industry and user groups, 228–230
  the importance of industry and user
    groups, 228
  where to look for industry and user
    groups, 228–230
    associations and national groups,
      229–230
      Association of American
        Geographers (AAG), 229

Canadian Association of
  Geographers (CAG), 229
Federal Geographic Data
  Committee (FGDC), 229
Geospatial Information and
  Technology Association (GITA),
  229–230
National States Geographic
  Information Council (NSGIC),
  230
Open GIS Consortium (OGC), 230
Urban and Regional Information
  Systems Association (URISA),
  230
  how to form an industry or user
    group, 230
  resources, 228–229
    Geo Community, 229
    GIS café, 228
  vendors, 229
    Autodesk User Group
      International, 229
    ESRI, 229
    Integraph, 229
    MapInfo, 229
Inertial navagation system (INS), 93–94
Information redundancy, 60
Information technology (IT), 50, 54–55
Integrated options, 199–200
Integration requirements and options,
  201–208
  integration options, 202–208
    common spatial database, 205–206
    customized integration, 203
    middleware, 204–205
    periodic data update, 202–203
    preintegrated GIS extension, 206–207
    preintegrated suites and connectors,
      207
    publish and subscribe, 207–208
  requirements for integration, 201–202
    communications, 201
    compatible format and range, 201
    design, 202
    synchronization, 201
    referential integrity, 201–202
Intelligence, 39
Intergraph GeoSpatial Users Community
  (IGUC), 229
Intergraph MGE, 223
Investigate the procurement options,
  74–75

L

LANDSAT, 9
Language barrier, 139–140